ourism

Charles Seale-Hayne Library
University of Plymouth
(01752) 588 588
LibraryandITenquiries@plymouth.ac.uk

MIX
Paper from
responsible sources
FSC® C014540
www.fsc.org

ASPECTS OF TOURISM
Series Editors: Chris Cooper, *Leeds Beckett University, UK,*
C. Michael Hall, *University of Canterbury, New Zealand* and
Dallen J. Timothy, *Arizona State University, USA.*

Aspects of Tourism is an innovative, multifaceted series, which comprises authoritative reference handbooks on global tourism regions, research volumes, texts and monographs. It is designed to provide readers with the latest thinking on tourism worldwide and in so doing will push back the frontiers of tourism knowledge. The series also introduces a new generation of international tourism authors writing on leading edge topics.

The volumes are authoritative, readable and user-friendly, providing accessible sources for further research. Books in the series are commissioned to probe the relationship between tourism and cognate subject areas such as strategy, development, retailing, sport and environmental studies. The publisher and series editors welcome proposals from writers with projects on the above topics.

All books in this series are externally peer-reviewed.

Full details of all the books in this series and of all our other publications can be found on http://www.channelviewpublications.com, or by writing to Channel View Publications, St Nicholas House, 31-34 High Street, Bristol BS1 2AW, UK.

ASPECTS OF TOURISM: 86

Brexit and Tourism

Process, Impacts and Non-Policy

Derek Hall

CHANNEL VIEW PUBLICATIONS
Bristol • Blue Ridge Summit

DOI https://doi.org/10.21832/HALL7123
Library of Congress Cataloging in Publication Data
A catalog record for this book is available from the Library of Congress.
Names: Hall, Derek R., author.
Title: Brexit and Tourism: Process, Impacts and Non-Policy/Derek Hall.
Description: Bristol, UK; Blue Ridge Summit, PA: Channel View
 Publications, 2020. | Series: Aspects of tourism; 86 | Includes
 bibliographical references and index. | Summary: 'This book offers a
 multidisciplinary, holistic appraisal of the implications of the UK's
 withdrawal from the European Union (EU) for tourism and related
 mobilities. It attempts to look beyond the short- to medium-term
 consequences of these processes for both the UK and the EU' – Provided
 by publisher.
Identifiers: LCCN 2019029541 (print) | LCCN 2019029542 (ebook) | ISBN
 9781845417116 (paperback) | ISBN 9781845417123 (hardback) | ISBN
 9781845417130 (pdf) | ISBN 9781845417147 (epub) | ISBN 9781845417154
 (kindle edition) Subjects: LCSH: Tourism – Great Britain. | Tourism – European
 Union countries. | European Union – Great Britain.
Classification: LCC G155.G7 H35 2020 (print) | LCC G155.G7 (ebook) | DDC
 338.4/7914 – dc23
LC record available at https://lccn.loc.gov/2019029541
LC ebook record available at https://lccn.loc.gov/2019029542

British Library Cataloguing in Publication Data
A catalogue entry for this book is available from the British Library.

ISBN-13: 978-1-84541-712-3 (hbk)
ISBN-13: 978-1-84541-711-6 (pbk)

Channel View Publications
UK: St Nicholas House, 31-34 High Street, Bristol, BS1 2AW, UK.
USA: NBN, Blue Ridge Summit, PA, USA.

Website: www.channelviewpublications.com
Twitter: Channel_View
Facebook: https://www.facebook.com/channelviewpublications
Blog: www.channelviewpublications.wordpress.com

Typeset by Riverside Publishing Solutions.
Printed and bound in the UK by Short Run Press Ltd.
Printed and bound in the US by NBN.

Contents

Section D Global Britain?

Figures, Tables, Boxes

Figures

All photographs are by Derek Hall.

Tables

Boxes

Abbreviations

ABTA	Association of British Travel Agents
ACML	Anti-Common Market League
AELP	Association of Employment and Learning Providers
AHDB	Agriculture and Horticulture Development Board
AOC	Air Operating Certificate
APD	Air Passenger Duty
APEC	Asia-Pacific Economic Cooperation
APPG	All-Party Parliamentary Group
ASEAN	Association of South East Asian Nations
ATA	air transport agreement
B&B	bed and breakfast
BASA	bilateral aviation safety agreement
BCAB	Border Communities Against Brexit
BCC	British Chambers of Commerce
Benelux	Belgium, Netherlands and Luxembourg
BiE	British in Europe
bn	billion
BRIC	Brazil, Russia, India and China
BTA	British Tourist Authority
CAN	Community and Ancillary Seller Notice
CAP	Common Agricultural Policy
CBI	Confederation of British Industry
CCS	Coastal Communities Fund
CCTV	closed-circuit television
CEE	Central and Eastern Europe
CEZ	coastal enterprise zone
CFSP	Common Foreign and Security Policy
CI	Channel Islands
CIPD	Chartered Institute of Personnel and Development
COSME	(EU) Programme for the Competitiveness of Enterprises and SMEs
CPRE	Campaign to Protect Rural England

CPT	Confederation of Passenger Transport
CSDP	(EU) Common Security and Defence Policy
CTA	Common Travel Area
$	(US) dollar
DBEIS	(UK) Department for Business, Energy and Industrial Strategy
DCLG	(UK) Department for Communities and Local Government
DCMS	(UK) Department for Digital, Culture, Media and Sport
DECC	(UK) Department for Energy and Climate Change
DEFRA	(UK) Department for the Environment, Food and Rural Affairs
DExEU	(UK) Department for Exiting the European Union
DfE	(UK) Department for Education
DftE	Department for the Economy, Northern Ireland
DfID	(UK) Department for International Development
DMO	destination management organisation
DUP	Democratic Unionist Party (Northern Ireland)
€	euro
EAFRD	European Agricultural Fund for Rural Development
EAGGF	European Agricultural Guidance and Guarantee Fund
EaSI	(EU) Programme for Employment and Social Innovation
EBacc	English Baccalaureate
EC	European Community; European Commission
ECB	European Central Bank
ECFR	European Council on Foreign Relations
ECHR	European Convention on Human Rights
ECJ	European Court of Justice
ECoC	European Capital of Culture
ECREU	Expat Citizens' Rights in the EU
ECSC	European Coal and Steel Community
ecu	European Currency Unit (forerunner of the euro)
EDF	European Development Fund
EEA	European Economic Area
EEC	European Economic Community
EEF	The Manufacturers' Organisation (formerly Engineering Employers' Federation)
EES	entry-exit system
EFTA	European Free Trade Association
EHIC	European Health Insurance Card
EIA	environmental impact assessment
ELT	English language teaching
EMS	European Monetary System
EMU	Economic Monetary Union

ENP	European Neighbourhood Policy
EPRS	European Parliamentary Research Service
ERDF	European Regional Development Fund
ERM	Exchange Rate Mechanism
ESF	European Social Fund
ESN	European Social Network
ESTA	Electronic System for Travel Authorisation
ETIAS	European Travel Information and Authorisation System
EU	European Union
Euratom	European Atomic Energy Community
EURES	European Employment Services
EU-ETS	EU Emissions Trading System
EY	Ernst & Young
EZ	enterprise zone
FDI	foreign direct investment
FIFA	*Fédération Internationale de Football Association*
GATT	General Agreement on Tariffs and Trade (forerunner of WTO)
GCE	General Certificate of Education
GCSE	General Certificate of Secondary Education
GDP	gross domestic product
GNP	gross national product
GVA	gross value-added
HEI	higher education institution
HESA	Higher Education Statistics Agency
HGV	heavy goods vehicle (lorry/truck)
HLF	Heritage Lottery Fund
HMO	house in multiple-occupation
IAG	International Airlines Group
ICCA	International Conference and Convention Association
ICT	information and communications technology
IFS	Institute for Fiscal Studies
INE	Instituto Nacional de Estadística (Spain)
IOC	International Olympic Committee
IoD	Institute of Directors
IoM	Isle of Man
IPPR	Institute for Public Policy Research
ISC	Independent Schools Council
JCQ	Joint Council for Qualifications

km	kilometre
KPI	key performance indicator
£	pound sterling
LEP	local enterprise partnership
LGV	light goods vehicle (van)
LSE	London School of Economics
m	million
MEP	Member of the European Parliament
MFF	multi-annual financial framework
MFL	modern foreign languages (teaching)
ml	mile
MNC	multi-national corporation
MP	Member of (UK) Parliament
MSE	mega sport event
MSP	Member of the Scottish Parliament
NATO	North Atlantic Treaty Organisation
NFFN	Nature Friendly Farming Network
NFU	National Farmers Union
NHS	National Health Service
NSMC	North South Ministerial Council (Ireland)
NT	National Trust
OAD	Overseas Association Decision
OCTA	Overseas Countries and Territories Association
OECD	Organisation for Economic Co-operation and Development (previously Organisation for European Co-operation)
Ofoc	Our Future Our Choice
ONS	Office for National Statistics
OSCE	Organisation for Security and Co-operation in Europe
PHV	private hire vehicle
PROGRESS	(EU) Programme for Employment and Social Solidarity
PTD	Package Travel Directive
PTRs	Package Travel Regulations
QMV	qualified majority voting
RBS	Royal Bank of Scotland
RDA	Regional Development Association
RFID	radio frequency identification
RTE	*Radio Telefis Eireann*

SEA	Single European Act
SEO	search engine optimisation
SGP	Stability and Growth Pact
SMEs	small- and medium-sized enterprises
STEM	science, technology, engineering and mathematics
TEU	Treaty for European Union
TIER 1-5	Points-based system for UK visa applicants
TiSA	Trade in Services Agreement
tn	trillion
TOMS	Tour Operator Margins Scheme
UK	United Kingdom of Great Britain and Northern Ireland
UKIP	United Kingdom Independence Party
UKOT	United Kingdom Overseas Territory
UKVI	UK Visas and Immigration
UN(O)	United Nations (Organization)
UNCTAD	United Nations Conference on Trade and Development
UNFCCC	UN Framework Convention on Climate Change
UNWTO	UN World Tourism Organization
US/USA	United States of America
V&A	Victoria and Albert (Museum, London)
VAT	value-added tax
VFR	visiting friends and relatives
WEU	Western European Union
WTM	World Travel Market
WTO	World Trade Organization

Preface

'I remain convinced that the European Union must be confronted
from *within*.'
Varoufakis, 2017: xi

This volume is very much about mobilities and immobilities. It is concerned with the unravelling consequences of societal division – in a time of imposed austerity – between the mobile (and perhaps presumptuous) and those (neglected) for whom spatial and social mobility is limited and for whom the mobility and apparent encroachment of 'otherness' is perceived as alien and (potentially) threatening.

The UK EU withdrawal debacle well illustrates the influences political decisions (or non-decisions) can exert on tourism and related mobilities, and, conversely, how the expressions of relative social and spatial mobility can influence politics. The UK's EU exit highlights the way in which tourism, as an object of study, is enmeshed within a wide range of social, economic, political and environmental considerations that may be closely and inextricably interrelated.

Partly as a consequence, this has not been one of the easiest books to put together. Not least because for much of its preparation period little appeared to be happening in the UK government's negotiations to leave the EU and thus there existed little more than a choice of speculative scenarios of the likely consequences. And towards the end of the book's preparation political confusion and absurdity actually heightened.

The propaganda word *Brexit* (see Hall, 1981) is not one that comes easily to the author's keyboard. It was incorporated into the title of this work on the understanding that despite an evident lack of clarity of what it might mean in practice, its intent appeared to be widely understood in principle. Except that ... being *Brexit* and not *UKexit* the term would appear to suggest an implicit exclusion of Northern Ireland ...

It is not a little ironic also that the single parent from whom the term is derived – *Grexit* – had not, at least at the time of writing, itself become a reality. Assuming, therefore, that by the time these words are read the UK has actually withdrawn from the EU, to mix already confused metaphors, the bastard child will have become the divorcee parent.

The purpose of this book is therefore to attempt to make some sense of the implications for tourism and related mobilities arising from the nature and processes of the UK's withdrawal from the European Union. To this end, the book is divided into four major sections with a total of 14 chapters.

Section A, *Context*, contains four chapters. Following a brief historical framework for the UK's (tourism) position in Europe (Chapter 1), Chapter 2 discusses the political context and logistics (if not logic) of the nature and outcome of the UK's 2016 EU referendum. Expressed through the EU's tourism role, the third chapter provides retrospect and prospect in addressing some of the challenges the EU faces without the UK. Chapter 4 offers a number of possible theoretical approaches to understanding the significance of *Brexit* for tourism.

In Section B, *Tourism Impacts*, Chapter 5 examines UK government and tourism business perspectives on likely impacts of the UK's EU withdrawal on domestic and international tourism. This naturally complements the following two chapters, addressing supply-side consequences of EU withdrawal for the sector (Chapter 6), and examining demand-side issues (Chapter 7).

Section C, *Implications*, begins with an evaluation of the likely environmental implications of EU withdrawal and their significance for tourism (Chapter 8). There then follow evaluations of three sets of 'inconvenient cross-border mobilities' relating to two of the three areas of contention that were required by the EU to be resolved by the UK during 2017 before substantive talks on a future relationship could begin. Significant for tourism-related mobilities, two of these three pre-requisite 'challenges' concerned (a) the question of the UK-Irish border and issues of accessibility, mobility and 'normality' across it (Chapter 9), and (b) the safeguarding of the rights of EU citizens living and working in the UK and of UK citizens living, working and retiring in other EU member states (Chapter 10). Virtually ignored in these earlier discussions and their media representation was the related situation of Gibraltar, which, although an overseas territory, had gained EU membership by virtue of being part of the United Kingdom, and was also exiting with the UK, despite overwhelming opposition to this in the territory's referendum vote (Chapter 11).

Section D, *Global Britain?* then attempts to look beyond the short- to medium-term consequences for UK tourism of EU withdrawal. Chapter 12 evaluates tourism's post-*Brexit* place in the reality of the Commonwealth's vaunted role for a 'Global Britain', while Chapter 13 examines the potential of the outbound Chinese market, including the role of student mobility, for 'a more outward looking' UK. The concluding chapter (14) takes *VisitBritain*'s marketing theme of the UK as a 'GREAT' global destination, and considers the reality of this in light of EU withdrawal policies and processes. While this evaluation brings the book to a close, it merely acts to open up a wider debate.

I would like to extend a sincere thanks to the four invited contributors to the book for their continued support, and, not least, for their timely and insightful pieces produced under somewhat uncertain circumstances. I should also like to thank Sarah Williams, Florence McClelland and the production team at Channel View.

Derek Hall

Seabank
Maidens
Ayrshire
Scotland

January and June 2019

Author Details

Derek Hall is former head of department and professor of regional development at the erstwhile Scottish Agricultural College, with particular interests in tourism and political geography. Pseudonymously, he is also a fiction writer.

Invited Contributors

Constantia Anastasiadou is a Reader in Tourism at Edinburgh Napier Business School, Edinburgh, Scotland, specialising primarily on tourism policy and planning and destination management. She has researched and written extensively on aspects of European integration and tourism.

C. Michael Hall is Professor in the School of Business, University of Canterbury, New Zealand; Docent, Department of Geography, University of Oulu, Finland and a Visiting Professor, School of Business and Economics, Linnaeus University, Kalmar, Sweden. He has published widely on tourism, regional development and environmental change.

Rong Huang is an Associate Professor in Tourism Marketing at the University of Plymouth. She has undertaken research into experiential tourism particularly in relation to Chinese international students and Chinese tourists. She has taught and researched into different areas of special interest tourism such as food tourism, tea tourism, literary tourism, film tourism and coastal tourism.

Lesley Roberts is a British expat living and working in Lucca near Florence. Formerly, she was a senior lecturer at Northumbria University and a researcher/consultant at the Scottish Agricultural College. She has published on rural tourism, tourism development in Central and Eastern Europe and the tourism consultancy process.

Section A

Context

1 The UK and 'Europe'

> *The study of the EU is filled with theories of European integration. Brexit confronts us with the need to theorise European disintegration ... [yet] Liberal inter-governmentalism draws in a mix of interests, institutions and ideas to highlight that Britain and the EU (especially Germany, France and other big states) are caught up in a deeply enmeshed set of interdependencies from which there is no easy escape whatever their leaders may want.*
>
> Oliver, 2015: 3

1.1 The UK-'Europe' Relationship

It is claimed (particularly by British authors) that it was Winston Churchill's 'United States of Europe' speech in 1946 that helped to crystallise thinking towards a time when all Europeans would see themselves as part of a peaceful whole (Danta & Hall, 2000). Certainly, such ideals of a united Europe were the inspiration that led, via the setting up of the Benelux customs union in 1948, to Jean Monnet and Robert Schuman's 1950 plan for an alliance between France and Germany that established the first instruments of European integration. These were the European Coal and Steel Community (ECSC), the European Atomic Energy Community (Euratom), and, most significantly, the European Economic Community (EEC), which came into effect in January 1958.

While the waning empire/colonies persisted through the 1950s, Britain remained aloof from (mainland) European cooperation. And when in 1961 the United Kingdom applied for EEC membership, it was twice vetoed by the French president Charles de Gaulle (in 1963 and 1967), not gaining accession until 1973. Although, or perhaps because of, having sought refuge in England during much of the Second World War, de Gaulle recognised that because the UK had avoided the experience of both wartime occupation and defeat, the country did not appear to embrace those fundamental desires for the sharing of sovereignty with other European peoples that drove Monnet and Schuman. As Clement Attlee put it in 1957, 'We're semi-detached'.[1]

The insular UK tended to be dominated by a drive for mercantilism coupled with a misplaced sense of superiority that is still evident today, perhaps even magnified, in some circles. The Commonwealth, and a perceived 'special relationship' with the United States, diffused the country's priorities.

> Hence the focus on the economic aspects of integration that has been common among British politicians and has restricted their ability to play an influential and constructive part in some of the most significant developments ... [including t]he EU's potential contribution to making the world a safer place in fields such as climate change and peacekeeping. (Pinder & Usherwood, 2013: 3)

1.2 The 'European Process'

While the UK looked on until 1973, a number of complementary bodies were established alongside the EEC to further its goals (Box 1.1).

The main institution's name has changed over time, to some extent reflecting the evolving nature of its role and objectives. The EEC (also

Box 1.1 Complementary European institutions

European Commission: to handle the bureaucracy involved in running the institution

Council of Ministers: as the main decision-making body with responsibility for broad policy formation and implementation, with a Council Presidency rotating every six months

European Parliament: to provide a democratic forum for debate and to operate as a watchdog

European Court of Justice: to hear cases involving member states

European Council: of heads of states meeting at least twice yearly to discuss issues before the Community (and later Union)

Committee of the Regions: to bring local concerns to the attention of the Council.

Most functions have been carried out in Brussels, but with Luxembourg and Strasbourg also as important centres of activity.

The Community set out to work closely with other European organisations, such as: Organisation for European Co-operation (later the Organisation for Economic Co-operation and Development: **OECD**)

North Atlantic Treaty Organisation (**NATO**, founded 1949)

Western European Union (**WEU**, 1954)

Organisation for Security and Co-operation in Europe (**OSCE**, 1973).

Sources: Danta and Hall, 2000: 6; Hall *et al.*, 2006: 6.

known in the UK as the 'Common Market' after a customs union came into effect in 1968), Euratom, and the ECSC combined during the 1980s to form the 'European Community' (EC). In 1993 the Community underwent a major re-organisation which subsumed, but did not replace, the EC; the overarching body now becoming the 'European Union' (EU).

Two broad courses of action that have characterised the EEC, EC and EU have been those of 'deepening' and 'widening'. 'Deepening' (Box 1.2) has involved mechanisms aimed at bringing member countries into closer economic, political, administrative, and security 'alignment'. Again, it is notable that strong foundations for this were established prior to the UK's accession.

Box 1.2 Stages of 'deepening'

Treaty of Paris (1951) established the ECSC
Treaties of Rome (1957) establishing the EEC and Euratom
Common Agricultural Policy (CAP, 1962)
Completion of a Customs Union (1968)
European Council (1975)
European Monetary System (EMS, 1979)
Single European Act (1986), which set the goal of creating a single market by 1993
Schengen Convention (1990) proposed the abolition of internal border controls and a common visa policy, creating the Schengen Area in March 1995. Formally, the agreements and rules adopted were separate from EC structures as they did not meet with consensus among EC member states.
Treaty on European Union (TEU), also known as the Maastricht Treaty (1993), set the goals of achieving monetary union by 1999; new common economic policies; European citizenship; common foreign and security policy; and common policy on internal security. Indeed, Maastricht's three main pillars of economic and monetary union (EMU), commitment to a common foreign and security policy (CFSP), and internal security, including trans-border mobility and crime, changed the 'architecture' of the Union to which new applicants would seek accession.
Treaty of Amsterdam (1997), covering aspects of justice and home affairs.
Common monetary policy and single currency, the euro (1999).

Although the Single European Act had marked a decisive shift towards majority voting in the Council, and the Maastricht Treaty transferred some powers to the European Parliament, decision-making remained something of a compromise between inter-governmentalism and cooperative federalism.

Sources: Danta and Hall, 2000: 7; Hall *et al.,* 2006: 6–7.

Although set out in the Treaty of Rome, the EU's 'four freedoms',[2] covering the free movement of goods, services, capital and people, were extended under the 'internal market' rules introduced by the Single European Act. Such freedoms, however, may be seen to be potentially in conflict with the EU's environmental goals of sustainability.

The 1993 Copenhagen criteria for accession included the development and sustaining of a functioning market economy, the capacity to accommodate competitive pressure and market forces within the Union, the development of democracy, the rule of law, human rights and the protection of minorities, and the ability of countries to accept the obligations of the Union which included some 20,000 laws and regulations of the *acquis* (Hall, 2017: 31), laws and regulations from which any member foolish enough to consider leaving the EU would need to disentangle themselves.

The second goal, 'widening' (Figure 1.1), has been pursued through a series of geographical enlargements, although from 1958 until 1973 the EEC consisted solely of the six members of the ECSC: Belgium, the Netherlands and Luxembourg (the former Benelux members), France, Germany and Italy.

Certain 'holdover' countries have remained outside the Union for differing reasons. Norway applied for membership in 1962 and negotiations were concluded 10 years later, but referenda held in the country in 1972 and 1994 failed to return a sufficient majority in favour of entry, although the country has remained within the EEA. Switzerland's desire for neutrality (and financial secrecy) has outweighed any perceived advantages of membership. The country rejected membership of the European Economic Area (EEA) and thus the possibility of EU entry in 1995. Iceland has also shown no interest in membership. Greenland (*Kalaallit Nunaat*) is one 'country' to have actually withdrawn from the European Union, in 1985, having become a self-governing overseas administrative division of Denmark in 1979[3] (Abbott, 2016). By contrast, Turkey first applied for membership in 1987.

UK public debate on the nature and significance of such experiences as possible pathways outside the EU tended to be limited and superficial (see Table 2.2).

Some of the most significant outcomes of the combination of deepening and widening processes have related to the encouragement of the free mobility of people within the EU. This has been reflected not only in tourism and leisure travel, but in education and employment mobility. The migration of relatively cheap labour from east to west has represented both a concern and a significant boost for some national economies, not least that of the UK, especially in terms of providing a flexible workforce for many elements of the tourism and hospitality sectors (Chapter 6) and seasonal labour in horticulture where local supply has appeared insufficient.

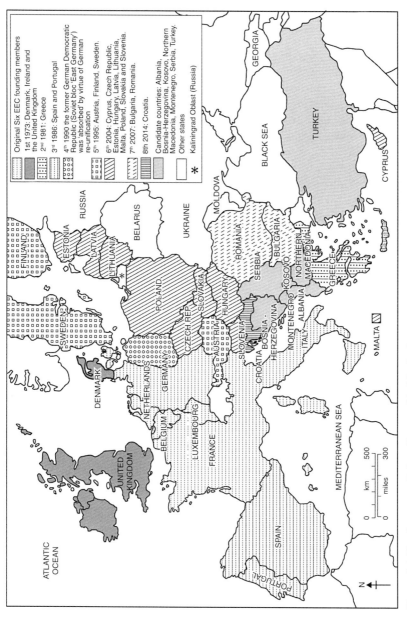

Figure 1.1 The European Union: Stages of 'widening'

Original Six EEC founding members

1st 1973: Denmark, Ireland and the United Kingdom

2nd 1981: Greece

3rd 1986: Spain and Portugal

4th 1990 the former German Democratic Republic (Soviet bloc 'East Germany') was 'absorbed' by virtue of German re-unification

5th 1995: Austria, Finland, Sweden.

6th 2004: Cyprus, Czech Republic, Estonia, Hungary, Latvia, Lithuania, Malta, Poland, Slovakia and Slovenia.

7th 2007: Bulgaria, Romania.

8th 2014: Croatia.

Candidate countries: Albania, Bosnia-Herzegovina, Kosovo, Northern Macedonia, Montenegro, Serbia, Turkey.

* Kaliningrad Oblast (Russia)

Other states

As a concomitant to the removal of internal barriers to the flow of people, goods and ideas within the EU, there has been a requirement to tighten the external borders of the Union, a dimension at the heart of the apparent mutual incomprehension between the EU and the UK government concerning the UK-Republic of Ireland border (Chapter 9). Crossing the external borders of the EU to enter one member state in effect provides legal entry to all EU member states, internal differentiation between Schengen zone and non-Schengen members notwithstanding. With tourism as an essentially national competence, EU mobility strategies have nonetheless emphasised the range of policy areas that impact upon tourism, including those for SME growth, supply chains, welfare and environmental sustainability (see Chapter 3).

Within this context, until relatively recently tourism had '... registered great difficulties in obtaining its legal and political recognition' within the EEC/EC/EU frame (Rita, 2000: 434–435). Rita attributed this situation to three apparently interrelated factors:

(i)   a misunderstanding of tourism's breadth coupled to and resulting from an inadequate evaluation of its economic, social and cultural significance;
(ii) a political under-assessment of tourism activity and its potential for Europe's future; and
(iii) obstacles within the Commission, successive presidencies and certain member states. (Rita, 2000: 435)

Certainly the EU has been significant in providing advisory frameworks and guidelines for the tourism sector within Europe, and perhaps most notably in the area of consumer protection. One of the better known examples of this being the 1993 Directive on Package Travel (updated 2013) which made it illegal for travel organisers and/or retailers to fail to provide their clients with necessary information on health needs.

While the European Commission has recognised tourism as one of the most significant activities in the EU in terms of economic benefits (employment, income and wealth creation in local communities), and social benefits (as a framework for the stewardship of distinctive cultures and environments), tourism has been denied a guiding policy that could be developed into a European model (Anastasiadou, 2011; Halkier, 2010; Manente et al., 2013).

As Estol and Font (2016) highlighted, tourism has remained a tool or 'common action' – in practice a means to an end – for such grander policy objectives as convergence to the Single Market (Anastasiadou, 2008b; Aykin, 2012), equity and social cohesion, sustainable development (Anastasiadou, 2011) and competitiveness. Tourism has been employed to help facilitate supranational competences (Panyik & Anastasiadou, 2013), while the distribution of tourism-related competences within member states has remained un-harmonised.

The nature and role of EU policymaking on tourism is developed further in Chapter 3.

1.3 Interconnectedness

This concluding section summarises interrelationships potentially disrupted with the UK's withdrawal from the EU, and thus heralds the focus of the rest of this volume.

The UK's EU budget contribution in 2017 was £16.9bn, but £8.1bn net (around 0.6% of UK GDP) after Margaret Thatcher's 1985 rebate of £4.8bn is deducted in advance, not forgetting the EU funds disbursed in the UK: £4.1bn, and the £1bn worth of EU competitive contracts won by British firms and bodies. The UK is the EU's third largest budget contributor, and, unsurprisingly, the 14 member states who are net budget recipients preferred that the UK remained in the Union (White, 2017).

Hitherto, approaching half of total UK exports has been accounted for by the EU, making it the single most important market for the UK. Over half of total UK imports have also come from the EU. These have made up around 10% of total EU exports, resulting in the UK being the third largest importer of EU goods and services after Germany and France. UK trade in services has consistently run a surplus since 2005.

In 2015 EU countries contributed 48% of the total UK inflow of FDI, while they accounted for 40% of UK FDI outflow. The UK was then the most important FDI destination in Europe, receiving $35bn in capital investment from both within and beyond the EU. However, following the EU referendum, this figure fell to $15bn in 2017, alongside an overall reduction of FDI into the EU of 42% (UNCTAD, 2018). In 2016, between 47% (England) and 74% (Northern Ireland) of the migrant workforce in the UK were EU citizens (Table 6.2; People 1st, 2017a). EU citizens made up 5.9% of the total UK workforce. In 2014, 27% of all UK emigrants migrated to EU countries for work-related engagements (Deloitte, 2016).

The UK and EU tourism sectors are highly interdependent, spatially and structurally. The EU is the main destination for UK tourists overseas, with approaching 54m or almost 75% of UK outbound visitors travelling to other EU destinations (Table 7.2). In terms of numbers the EU is the main source of tourists coming to the UK, with almost 25m or 65% of overseas visitors to the UK (Table 7.7). Visitors from the EU contribute around £10bn to the UK economy annually (Table 7.9), while outbound tourism to the EU generates an estimated £22.4bn for the UK economy

Table 1.1 UK's top five EU travel markets and destinations, 2018

Number of inbound visits (in millions)		Number of outbound visits (in millions)	
1. France	3.69	1. Spain	15.62
2. Germany	3.26	2. France	8.56
3. Ireland	2.78	3. Italy	4.33
4. Spain	2.53	4. Ireland	3.22
5. Netherlands	1.95	5. Portugal	2.82

Source: ONS, 2019b: Tables 2.10, 3.10. See also Tables 7.8 and 7.4.

(Tourism Alliance, 2017). The UK's top five EU travel destinations and markets in 2018 are shown in Table 1.1.

Since 2008 French residents have headed the table for the most visits to the UK and 2018 was no exception. For an even longer period Spain has been the most popular destination for UK travellers. In terms of crude numbers, in 2018 8 of the top 10 and 14 of the top 20 visitor source countries to the UK were from the EU (Table 7.8). In terms of spend, those numbers fell to 6 and 11 respectively (Table 7.10). For UK outbound traveller numbers, 9 EU destinations featured in the top 10 and 13 in the top 20 (Table 7.4); for visitor spend the figures were 9 and 10 respectively (Table 7.5).

Some 68% of foreign business visits from the UK are to EU destinations, while 73% of overseas business visitors to the UK are from EU countries. Maritime and air passenger transport are particularly reliant on passenger flows between the UK and other EU member states. By sea, EU countries contribute over 85% of the total passenger traffic to UK ports. In aviation, travel to EU destinations accounts for 64% of the UK outbound passenger flow (Reckless Agency, 2017). An 85% increase in passenger numbers at UK airports between 1996 and 2016 appeared largely due to the entry and rapid growth of low-cost carriers following the introduction of the EU's 'open skies' policies.[4]

UK and European travel, accommodation and hospitality companies have operated seamlessly on all sides of the Channel/North Sea/Irish Sea.

One of the main facilitators for this high level of interdependence has been the harmonised single market. This framework of agreements, rules and regulations has encouraged tourism flows between the UK and the rest of Europe, allowing the UK tourism sector to grow to become one of the largest in the world and a key component of the UK economy (Tourism Alliance, 2017: 4).

The nature of the UK's withdrawal from the EU had the potential to significantly disrupt this process.

Notes

(1) At a meeting in Clarens, Switzerland, organised by the American Friends Service Committee, August 1957 (Marshall, 2016: 454).

(2) Not to be confused with Franklin D. Roosevelt's 'four freedoms', when he insisted that people in all nations of the world shared Americans' entitlement to: the freedom of speech and expression, the freedom to worship God in his own way, freedom from want, and freedom from fear.

(3) Greenland remains associated to the EU as an Overseas Territory.

(4) It should also be noted that over this period the number of UK citizens taking cruises increased four-fold.

2 Imbroglio

This chapter attempts briefly to outline the political process that brought about the UK's withdrawal from the European Union.

2.1 It's My Party …

The two major UK political parties had long been internally divided over the country's role in relation to the European Union and its predecessors. In recent decades, the Conservative Party in particular had been harried from within by a heterogeneous alliance of 'Eurosceptics'. However, acting as a populist single-issue party, with an essentially single populist figurehead, the United Kingdom Independence Party (UKIP), originating in name from 1993, gained popular support in marginalised areas, particularly within England, in the 2013 local elections and 2014 European elections. UKIP's relative success encouraged the 'Eurosceptic' element of the Conservative Party to be more vigorous and far-reaching in its aspirations. In the 2015 general election UKIP gained 3.8m votes across the UK, with 12.6% of the votes cast, albeit with only one Member of Parliament to show for it.

Panicked and bullied into action,[1] the Conservative Party leadership had drafted a European Union (Referendum) Bill in May 2013, providing for a referendum on the question of the UK's position within the EU to be held by the end of 2017. The party's 2015 general election manifesto 'reiterated a commitment to an in/out referendum' on EU membership *on renegotiated terms before* 2017 (Uberoi, 2015).

Such renegotiated terms of membership that were belatedly secured by prime minister Cameron at a European Council meeting in the spring of 2016, constituted a 'set of arrangements' that amounted to:

- the UK not taking part in further political integration in the EU;
- any future UK contributions toward eurozone monetary measures being voluntary;
- some benefits paid to immigrants being limited;
- a target being set for reducing the 'burden' of regulation; and
- a commitment to strengthen the internal market.

This narrow range of symbolic concessions had been gained that was unlikely to harm significantly other member states' interests

(Kroll & Leuffen, 2016). While most of these revised terms would apply to the UK economy as a whole, they would result in travel and tourism sector businesses experiencing a number of changes in bureaucratic requirements. Crucially, however, the UK sector would continue to be able to influence EU policymaking. 'It was an expandable bridgehead; (Marshall, 2016: 454), but there followed an almost complete absence of any serious examination of the agreement, and a referendum was called.

In the UK's EU referendum of June 2016, the choice was between a known, if evolving, relationship between the UK and the EU, and the unknown position of the UK being outside the EU. David Cameron famously intended the EU referendum to 'settle the issue for a generation' (Hyde, 2017: 34). That the outcome of remaining in the EU had been considered a foregone conclusion by those in power blinded them to the potentially momentous and damaging contrary outcome, for which they were clearly unprepared. There was no deliberative process to inform public opinion, neither was effort made by government to indicate what type of 'Leave' it would choose if the vote went that way.

The Referendum Bill had not proposed a voting threshold (Uberoi, 2015: 26): only in relation to the Scotland and Wales devolution referenda of 1979 had thresholds ever been invoked in the UK for such mechanisms. These were provided for in the Scotland Act 1978 and the Wales Act 1978, following a backbench amendment requiring a minimum of 40% 'of the persons entitled to vote in the referendum' voting 'yes' (Gay & Horton, 2011: 3).

Referenda, some argue, are not truly democratic. They challenge the assumptions upon which a system of democratic representative government is based (Mayall, 2016). They are a blunt instrument: they can be employed to conflate complicated political arguments into a simple – too simple – yes/no final answer. Referenda can act to heighten emotions and deflect reasoned deliberations. Even Margaret Thatcher referred to them as 'a device of dictators and demagogues' (Alibhai-Brown, 2017: 10). Importantly, referenda can effectively weaken parliamentary scrutiny as politicians acquiesce before a 'popular' vote.

Thus, the complex UK-EU relationship that had evolved over 43 years was reduced to a simple remain/leave choice. And for such a constitutionally important issue to be decided upon a simple majority vote was clearly inappropriate and injudicious.

The manner in which the subsequent campaign was conducted by 'Leave' proponents generated considerable passion amongst many people who were not habitual voters, particularly on the issue of immigration. The 'Remain' campaign was dominated by the government, thereby denying opportunities for cross-party endeavour. As a consequence, elements of the electorate appeared to consider the vote as a verdict on the government.

'Leave' proponents in the major parties distanced themselves from UKIP's neo-racist propaganda and shifted to a more subtle psychological approach. The repeated phrase 'Take back control' took centre stage. It indicated the recognition of about 650,000 swing voters (Oliver, 2017: 403) who felt neglected and unheard, were concerned about National Health Service (NHS) funding and immigration, and who were located particularly in de-industrialised and economically deprived regions.

Such fears and anxieties that were played to by 'Vote Leave' and its 'Take back control' *mantra*, were coupled to the wildly false claim, infamously plastered over a bright red 'battle bus', that £350m a week saved by being outside the EU would go to the NHS (Greer, 2017).

Later, both the Electoral Commission (2018) and the High Court in London (Ramsay, 2018) ruled that, out of the complicated (financial) relationships between the various 'Leave' campaigns, there had been breaches of electoral law (Cadwalladr, 2018; Geoghegan, 2018). The matter was referred to the police.

2.2 Referendum Outcome

Viewed as one symptom of a continuing organic crisis of the British state and society (Jessop, 2017: 133), the referendum motivated more than 30m UK nationals to vote; with a majority in favour of leaving the EU at 51.9% (representing 37% of the enfranchised electorate[2]) to 48.1% wishing to remain, on a turnout of 71.8% of the total electorate.

Viewed superficially, England voted for *Brexit*, by 53.4% to 46.6%. Wales also voted to leave by 52.5% to 47.5%. Majorities in both Scotland and Northern Ireland supported staying in the EU by 62% to 38% (Curtice, 2017) and 55.8% to 44.2% (McCann & Hainsworth, 2017) respectively. A majority in all 32 local authority areas in Scotland voted to remain (MacKenzie, 2016).

Alongside Scotland and Northern Ireland, London and most English metropolitan areas with universities voted to remain within the EU (Bristow *et al.*, 2016). Young people overwhelmingly voted to remain, but had to face the consequences of withdrawal in a future where many of those who voted to 'Leave' would no longer be around.[3]

The unforeseen outcome appeared to have resulted from a combination of factors that varied regionally and demographically (e.g. Jablonowski *et al.*, 2018; Kiefel *et al.*, 2018). These included elitist ignorance and complacency; a sense of voicelessness among large proportions of the population in de-industrialised regions concerning austerity issues unrelated to the vote in question; a background apprehension towards 'outsiders', largely heightened by malevolently reactionary elements of the UK press coupled to wilfully mendacious politicians seeking to position themselves for future leadership roles; and a well-orchestrated encapsulation of a number of mythologies rolled into the totalitarian-like repeated *mantra* 'Take back control'.

One intriguing potential contributory factor in the vote may have been the wording of the referendum ballot paper itself: while *Leave* is an ancient word of everyday intercourse, *Remain* is both more recent and more formal (compared to *Stay*) (Trudgill, 2017) (Box 2.1). Further, there is a certain sense of heroic dogged persistence attached to someone who is a *stayer* that is absent in the far more passive, and perhaps even 'left behind' *remainer*.

Is it totally impossible that the result of the EU referendum could have been different if the ballot paper had asked us, not if we wanted to remain in the EU, but if we wanted to stay? After all, the wording for the 1975 referendum was: 'Do you think that the United Kingdom should stay in the European Community (the Common Market)?' The result back then was a 67 percent 'yes' vote. (Trudgill, 2017: 31)

Box 2.1 *Leave* and *Remain*

Leave	Ancient Germanic word, brought by the Anglo-Saxons 350–550AD, thus part of English ever since it was a language. Has been a part of natural, everyday speech.
Remain	Not used in English until the fifteenth century. Originally only employed by Norman French overlords. Therefore has elitist connotations and generally absent from normal, everyday conversation.

Source: Trudgill, 2017.

The UK referendum vote and the election of DJ Trump as US President, within five months of each other, could be seen as proxies for a whole cluster of frustrations, largely, but not wholly, arising from the financial crisis and continuing austerity programmes. Yet in the months, and years, following that referendum, there was severely limited government effort to address those underlying issues.

And yet ... a *YouGov* poll held in the summer of 2017 found that 61% of 'Leave' voters believed economic damage was 'a price worth paying for Brexit', with 39% of 'Leave' respondents declaring that members of their family losing their jobs would be worth the cost in order to exit the EU. Significant age group differences were apparent (Table 2.1) (Wilkinson, 2017).

Table 2.1 *YouGov* poll on strength of *Brexit* feeling by age, August 2017

Percentage of *Leave* voters by age group agreeing with the below sentiments

	18–24	25–49	50–64	65+
Significant damage to the economy would be a price worth paying for leaving the EU	46	56	60	71
The loss of their own or their relatives' job is a price worth paying for leaving the EU	25	32	38	50

Source: Wilkinson, 2017: 11.

In Scotland, with a majority 'Remain' vote in every local authority area, proponents of Scottish independence suggested that a second independence referendum could be held after EU exit negotiations were completed. That raised the question as to how far some independence supporters may have voted tactically to leave the EU in order to precipitate this.

2.3 Invoking Article 50

An immediate outcome of the referendum vote was a new prime minister – Theresa May. David Cameron resigned on the morning after the vote. Like her predecessor, May had been opposed to the UK leaving the EU but played a low-key role in the campaign. She became prime minister without facing a full Conservative leadership contest after her key rivals – 'Leave' proponents – withdrew.

Building on Cameron's Referendum Bill (Uberoi, 2015), the May government's white paper (UK Government, 2017b) of 2 February, aimed to present 'a clear vision' of what it was seeking to achieve in negotiating the UK's exit from, and new partnership with, the EU. The paper's internal contradictions and oversimplified options were evident in subsequent negotiations.

Significantly, the white paper did appear to support completion of the single market for services, given the importance of this sector as a whole for the UK economy, at around 80% of GDP, and for the tourism and hospitality sectors in particular.

Nine months after the referendum vote, on 29 March 2017 the UK prime minister invoked Article 50 of the Lisbon Treaty (Treaty for European Union: TEU). This was the only formal process for a member state to resign its EU membership, initiating up to two years of negotiations to agree the terms of withdrawal and a future relationship. If the two-year timeline elapsed, this could be extended with the unanimous agreement of the EU member states. If no extension was agreed, the exiting country could revert to a basic set of trade rules pending further negotiations outside of the Article 50 provisions.

As no member state had actually left the EU prior to this (Greenland notwithstanding), much uncertainty surrounded the manner in which the process might proceed. Such uncertainty prevailed considerably longer than envisaged, encouraging prolonged and divisive debate of possible alternative scenarios. As indicated in Table 2.2, most conventional models required a continuing contribution to the EU budget and compliance with a large proportion of EU regulations. Neither of which, a good proportion of the press and politicians supporting withdrawal would contend, was envisaged in the 'Leave' vote.

Some *Brexiters* appeared to relish the possibility of the UK leaving the EU with no trade deal, arguing that World Trade Organization (WTO) rules were adequate for the country's 'global' trading role.[4] A potential problem – one of the many – with this notional arrangement was that WTO provisions did not cover services.

The shorthand, if misleading terms of 'soft' and 'hard' *Brexit* came into popular use. A 'hard' *Brexit* seemed to apply to a 'full' UK withdrawal from the EU: from its single market and customs union, and from jurisdiction of the European Court of Justice (ECJ), continued membership of these three being 'red lines' that the prime minister vowed would not be crossed (Polak, 2017). As not necessarily the polar opposite, a 'soft' *Brexit* appeared to imply staying within or being 'harmonised' with any or all of those three components that the prime minister precipitately proscribed.

Given the relatively small majority of the UK population voting in favour of EU withdrawal and the balance of arguments both within and outside the government, prime minister May called a general election for 8 June 2017, almost a year after the referendum vote, and which, it was assumed, would provide a stronger mandate for proceeding with withdrawal.

The ruling Conservative Party's 20 point lead in opinion polls evaporated over an election campaign that was badly led (Frost, 2017). With strong support for the opposition Labour Party from young voters, the government lost its hitherto clear majority. Finding an extra £1bn for the Northern Ireland exchequer, the Conservatives managed to recruit the ten-vote support of that province's socially conservative, pro-*Brexit* Democratic Unionist Party, to sustain their minority government. This would subsequently exacerbate complications over the future nature of the UK-Ireland relationship and border (Chapter 9).

Although the government produced briefing papers on the costs and implications of various exit options, it was left to others to draw meaningful comparative analyses. Portes (2018), for example, interrogated the government's own impact studies on three different exit scenarios (House of Commons, 2018) together with the government's preferred option, a 'bespoke' deal, employing data from the official assessments together with details set out by the prime minister (Prime Minister's Office, 2018).

The basic characteristics of these four scenarios, alongside the status quo, are shown in Table 2.3.

Table 2.2 Possible alternative scenarios to EU membership

		Possible scenarios and existing subscribers			
		EU MEMBER 28 European Member Nations	EUROPEAN ECONOMIC AREA: Norway, Lichtenstein, Iceland	FREE TRADE AGREEMENTS (EFTA) e.g. Switzerland	MOST FAVOURED NATION e.g. Australia
	Free movement of goods, services and capital	+	+	+	–
Existing EU Obligations	Free to negotiate trade deals and set tariff levels with non-EU countries	–	+	+	+
	EU laws and regulations — Influence	+	slight/indirect	–	–
	EU laws and regulations — Compliance	+	+	+	–
	Fiscal contributions	+	+	+	–
		Looser ties with EU >>>			

Source: Deloitte, 2016: 7.

Table 2.3 Characteristics of UK government's EU exit scenarios

	Status quo	EEA	Bespoke FTA (Government's preferred deal)	Average FTA	'Smooth' No deal (WTO)
Customs non-tariff barriers (EU)	None	High	None – high	High	High
Behind border non-tariff barriers (EU)	None	Low	Some – medium	Material	High
Tariffs (EU)	None	None	None – low	Low-material for agri-food	High
Rest of the World Trade	Constrained	Material potential	Seek maximum flexibility	Material potential	Material potential
Regulation divergence (UK choice)	Constrained	Limited potential	Seek maximum flexibility	Material potential	Material potential
EU migration (UK choice)	Continued labour mobility	Continued labour mobility	Range of policy choices	Flexible labour mobility	Strict labour mobility
Government estimate of additional borrowing in 2033/34 compared with status quo	None	£17bn	No estimate	£57bn	£81bn

Key: FTA: free trade association
Source: House of Commons, 2018; Portes, 2018.

Table 2.4 Projected costs of the four exit scenarios

Scenarios	Net additional borrowing in 2033/34 (£bn)	Cost per week in 2033/34 (£)	Cost per week in 2018 prices (£) (2033/34 deflated to take account of 25% growth)	Cost per week in 2018 prices: ie cost as a % of the 2018 NHS budget (2033/34 deflated to take account of 25% growth)
EEA	17	327m	262m	9
FTA	57	1.1bn	877m	31
WTO	81	1.56bn	1.25bn	44
Bespoke deal	40	769m	615m	22

Source: Portes, 2018.

The projected costs of the four scenarios (Table 2.4) suggested that each would leave the UK poorer, costing the taxpayer hundreds of millions of pounds every week. The calculated 'bespoke' option would increase the cost of non-tariff barriers by £23bn above the status quo position. Other costs, including customs barriers, 'divorce' payments (see below section 2.6) and ongoing contributions would add another £38bn, while limits on free movement would cost a further £6bn. The Portes (2018) analysis nonetheless found that a 'bespoke' agreement could also generate an extra £27bn from such sources as customs revenue and EU budget savings.

This study was variously reported, both to confirm the worst fears about the costs of *Brexit* (e.g. Crerar, 2018), or to argue that the study's fourth option was wrong in citing maximum WTO tariff rates, when they would not necessarily be charged: 'just about every Brexit report makes this same mistake. Or, in at least in some cases, deliberate error. Correcting for it gives us … Brexit making us 3% richer' (Worstall, 2018).

2.4 Divorce and Diversion

As a condition for proceeding to talks on trade relations, the EU position required from the UK government the resolution of three critical issues:

- the level of payment the UK must make to cover the cost of the commitments it had made as an EU member state: the 'divorce' bill,
- the future nature of the UK-Irish external EU border (Chapter 9), and
- citizens' rights, relating to the position of EU citizens resident in the UK, and that of UK expatriates living in EU member states (Chapter 10).

Official talks for the process of the UK's withdrawal from its 44-year-old relationship with the EU began in 2017. As a consequence of the prime minister's invoking Article 50 of the Treaty for European Union, the UK would be destined to leave the EU at 23.00 UK time on 29 March 2019. The EU negotiator, Michel Barnier hoped to conclude talks by October 2018 in order to allow time for the European parliament to ratify the agreement.

While the EU required the UK to leave the Union ahead of European elections in May 2019, it was emphasised that the door remained open for the UK to change its mind and revoke its 'intention' to withdraw. But the only way back in after a March 2019 withdrawal would be for the UK to re-apply for EU membership.

The UK government had been committed to paying into the EUs 'multi-annual financial framework' (MFF): the EU budget for 2014–2020. It also held commitments to other financial funds, aid programmes and to EU employee pensions. In some cases, withdrawal would mean the return of money to Britain, but the balance would be negative.

Only in November 2017 did the UK government acknowledge readiness to honour its financial commitments through the post-*Brexit* transition period, estimated at €40-45bn. Subsequently it became clear that payments would continue until around 2064, and indefinitely if the UK remained part of EU agencies and programmes (Comptroller and Auditor General, 2018).

But the entanglement of so many government departments in the withdrawal process appeared to divert attention, effort and resources away from a number of pressing domestic issues and overseas endeavours (e.g. Kentish, 2018), further eroding the country's image and soft power reserves abroad. *Brexit* was sucking the life out of public discourse.

The resigning chair of the government's social mobility commission observed that coping with *Brexit* meant that the government 'does not seem to have the necessary bandwidth to ensure the rhetoric of healing social division is matched with reality' (Savage, 2017: 1). The extent to which such an issue would have been addressed adequately without EU withdrawal, during a period of continuing government austerity and retrenchment, is, of course, little more than speculative.

> … most of our civil servants and diplomats are working on dismantling our EU membership rather than on maximising our influence around the world, – and paradoxically we are taking on thousands more to do so in the pursuit of less bureaucracy. (Powell, 2017: 25)

2.5 Transposing EU Law and Directives

An estimated 6.8% of primary UK legislation, and 14.1% of secondary legislation, stemmed from the EU (Miller, 2014; Miller, 2018). As part of the UK government's European Union (Withdrawal) Bill, EU law would be directly transposed into UK law. Section 11 of the

Bill stated that all powers currently exercised in Brussels would be repatriated to Westminster at the point of leaving the EU, even those covering devolved policy areas such as agriculture, the environment and tourism.

The Scottish government viewed this as a direct threat to the devolution settlement, estimating that 111 devolved policy areas were involved, and demanded that Section 11 be excised or drastically rewritten (Morris & Carell, 2018).

But as with the tardy appearance of the government's sector impact studies (Chapter 5), the excuse was offered that because delicate exit negotiations were being undertaken in Brussels, little information or indication of direction – including the redistribution of repatriated powers – could be released for fear of jeopardising the UK government's negotiating stance (Gordon, 2017). Subsequent amendments to the EU withdrawal bill saw the 'vast majority' of transposed powers being vested in Edinburgh, Cardiff and Belfast, although the UK government would retain control over certain aspects of legislation that were normally devolved but which 'clearly related to the UK as a whole', such as food labelling and hygiene rules (Sparrow, 2018).

And of course, after EU withdrawal, the UK would have little if any, influence over EU laws and rules as it would no longer participate in negotiations within the EU legislative process.

2.6 EU Position

In its *draft* guidelines on the future relationship, the EU argued that in signing up to a free trade agreement the UK government would be obliged to sign a commitment not to become a low-tax, low-regulation state that undercut the EU model. It pointed to Westminster's self-imposed 'red lines' – withdrawal from the customs union, single market and ECJ jurisdiction, no mechanisms differentiating Northern Ireland and the rest of the UK (Polak, 2017) – limiting the depth of any future partnership. With significance for tourism and mobilities, the EU draft included:

- a customs arrangement to minimise barriers to trade, albeit one with 'rules of origin' checks and other controls on borders;
- access to the EU market for UK seafood exporters in return for reciprocal rights for EU fleets to continue fishing in UK waters;
- an agreement on services, although the financial sector was not mentioned, and the UK would 'no longer share a common regulatory, supervisory, enforcement and judiciary framework'. The UK government subsequently indicated that it would not agree to the exclusion of financial services;
- an aviation accord;

- cooperation in EU programmes for 'research and innovation and of education and culture' in return for payments into the Brussels budget;
- calls for cooperation in defence, security and foreign policy;
- in the event of no agreement over the Irish border issue, Northern Ireland could remain in regulatory alignment with the EU: a 'backstop' position initially rejected by the UK government (and especially by its DUP support) (Johnson *et al.*, 2018).

In subsequently agreed terms for the April 2019–December 2020 transition period, EU freedom of movement rights would be sustained both for EU citizens in the UK and for UK citizens in the EU (thereby crossing one of the UK government's 'red lines').

2.7 'Chequers' White Paper and the UK-EU Withdrawal Agreement

Following a full Cabinet weekend summit during July 2018 at Chequers (the UK prime minister's official residence) to belatedly formulate (and agree) government *Brexit* policy, two key ministers (*Brexit* Secretary and Foreign Secretary) resigned, as a second major government white paper was produced (UK Government, 2018d).

The white paper suggested the UK was seeking to stay in a single market for goods but to have a much looser relationship for services: yet the two are not mutually exclusive (Morris, 2018). The interconnectivities between goods' and services' supply chains vary from company to company, many operating simultaneously. In the UK food sector, for example, the government's proposal would 'inject a fault line' between, on the one hand, farming and food processing, and on the other, food retail and service. Yet food service is by far the largest source of employment in the UK food chain, delivering more gross value added (29%) than other sub-sectors (Lang *et al.*, 2018).

A key part of the proposals would be a 'facilitated customs arrangement': a compromise to minimise border friction whereby the UK and EU would become a 'combined customs territory' for goods, in which the UK would apply the EU's tariffs and trade policies for goods intended for the bloc, with domestic tariffs imposed for goods heading to the UK, collecting EU tariffs on the EU's behalf: a procedure not undertaken on any significant scale anywhere else in the world.

This would involve discussions with the EU on areas such as a trusted trader schemes and arrangements for repaying excess tariffs. The government would explore options to use future [sic] advances in technology 'to streamline the process'.

Free movement of people from the EU to the UK would end after the transition period, but it would be necessary to recognise the depth of the relationship and close ties between the peoples of the UK and the EU (UK Government, 2018d). The UK would seek to still attract the 'best and brightest' from the EU. Citizens would be allowed to travel visa-free for tourism, temporary work/business trips, and for study.

The white paper also reiterated a desire to maintain 'high environmental standards' with a new statutory body to monitor this, but contained no details (UK Government, 2018d) (see Chapter 8).

Replete with contradictions – if there was no free movement of people how could the Irish border be frictionless? – the white paper was condemned as unworkable by critics in the UK, not least from within the government party. Nonetheless, it went forward as the basis for Brussels negotiations over the draft withdrawal agreement and accompanying future relationship document. When these agreements returned to Westminster political turmoil ensued. A withdrawal agreement needed to be ratified by the UK parliament before 30 March 2019. Three major attempts to pass the draft agreement as UK law were heavily defeated, and turmoil turned into stalemate.

As the EU would not renegotiate the agreement signed off by prime minister May, the UK was given a revised withdrawal date of 31 October 2019 on the understanding that British politics could resolve its differences and find majority support for the withdrawal agreement. This meant that the UK was required to participate in the May 2019 European elections, in which the ruling Conservative Party in the UK came fifth. Prime minister May announced her intention to resign, thereby prompting a leadership contest which would further eat into the time remaining before the revised departure date.

The continued possibility of an un-ratified, 'no-deal' *Brexit* further reduced the value of pound sterling. The European Commission had devised a contingency action plan for the EU27, in the event of such an outcome, covering fourteen sectors, including aviation, road transport and customs (European Commission, 2018g). While the UK Treasury allocated an extra £2bn to 25 government departments for the 2019–20 financial year to cover potential 'no-deal' consequences, a classified Cabinet Office paper set out emergency command, control and coordination structures – 'Operation Yellowhammer' – including 43 'local resilience forums', to cover a 'critical three-month' phase of potential economic and social disruption (Hopkins, 2019).

2.8 Conclusions

While the UK referendum vote represented a blow to the EU's geopolitical pride, it also acted to reduce internal criticism, holding up to potentially querulous member states just how difficult and painful breaking from the Union could be. This reinforced a reluctance to offer any concessions to the UK in withdrawal negotiations so as not to encourage others. Further, as a constraint on subsequent bilateral arrangements, none of the remaining 27 states would wish to see the UK appearing to be better off outside the EU than within it (Rankin, 2018). But in the 'rapidly evolving geometry of Europe' (Galbraith, 2017: 164) nothing was certain.

Strongly constraining the UK's EU withdrawal options, and thus the implications for tourism and mobilities, three critical decisions characterised the UK government's approach:

(i)   a far too early triggering of Article 50 in March 2017 (Table 2.5) that left only months to secure major agreements that might have required at least five years' work;

(ii) declaring a series of 'red lines' (noted above, section 2.3) that would not be crossed, severely limiting flexibility in negotiations; and

(iii) calling a general election for June 2017 (section 2.3 above) in the mistaken belief that it would endorse the outcome of the referendum and thereby increase the government's majority. Instead, the government lost its overall majority and found itself relying on the Northern Ireland Democratic Unionist Party (DUP) for support, a factor that was to prove critical regarding the border issue in Ireland.

Table 2.5 Framing *Brexit* for tourism impacts

Date	UK and the EU	Tourism-related significance
Pre-2016	Pre-referendum: non-€, non-Schengen	Being outside the Eurozone and Schengen imposed filters on the 'seamless' movement of people and goods between the UK and much of the rest of the EU
June 2016	Referendum vote: 51.9% for leaving the EU	Immediate fall in the value of sterling rendering the UK cheaper for visitors from Europe, while rendering overseas travel more expensive for UK travellers
March 2017	Triggering of Lisbon Treaty Article 50	Two years until the UK leaves the EU: many sectors argue that this leaves insufficient time for future planning
June 2017	General election: the UK government becomes dependent upon the DUP to sustain its parliamentary majority	Irish (external EU) border question and related issues were immediately complicated
September 2017	Prime minister May's Florence speech proposing a two year transition period	At least this would give more time for tourism-related businesses to plan ahead
December 2017	Preliminary negotiations on cost of exiting, citizens' rights and the question of the Irish border ultimately fudged	Continuing uncertainty: not least for UK expats in the EU and EU citizens in the UK (see Chapter 10)
March 2018	EU's draft guidelines on the future relationship: only '75%' agreed by UK government	Major unresolved issues and uncertainties remained

(Continued)

Table 2.5 (continued)

March 2018	Transition period agreed at 21 months	This provided clarity on freedom of movement issues at least until 31 December 2020
June 2018	Summit of EU leaders	By this time tourism and travel sector businesses needed certainty for planning the first post-*Brexit* summer season from 1 April 2019
July 2018	Government white paper: *The Future Relationship between the United Kingdom and the European Union*	Sought visa-free movement for tourists, temporary workers and students, and to maintain 'high environmental standards'.
		Other sections, notably on tariffs and customs arrangements, were seen as largely unworkable
July 2018	Heightened discussion of a possible no-deal exit in the wake of the heavily criticised white paper	The value of sterling fell further, especially against the US dollar, thereby rendering more expensive UK visits to the US as well as services and commodities such as fuel oil priced in dollars on world markets
November 2018	Negotiated withdrawal terms agreed by EU	Debate in UK parliament in December curtailed by the prime minister in the face of a likely rejection of agreement
January – March 2019	Further debate in UK parliament on negotiated withdrawal agreement	Agreement rejected three times by large majorities. All tabled alternatives also defeated
31 March 2019	Intended UK withdrawal date from the European Union	Postponed to 31 October by the EU in view of UK parliamentary deadlock. If the UK parliament voted in favour of the withdrawal agreement before 1 October, the UK would leave on the first day of the next calendar month
31 October 2019	Deadline for UK withdrawal from the European Union	See Chapters 3–13
1 November 2019-31 December 2020	Transition period: assuming withdrawal on 31 October	Does not apply to a 'no-deal' withdrawal.
		May be extended once by two years
1 January 2021	UK's final full EU departure after transition period	If transition period extended, the latest date would be 1 January 2023.

Sources: various, including: Government of the Netherlands, 2019.

If the UK government had begun fleshing out its ambitions earlier in the process – in practice obviated by the early invoking of Article 50 – it could have made possible both a more sympathetic hearing from the EU during negotiations, and, importantly, it would have provided UK tourism and travel businesses with a better understanding of, and more time to plan for, the practical nature and implications of the UK's EU withdrawal.

Although the UK government – not least through its *Tourism Action Plan* (DCMS, 2016) (Chapter 5) – expressed awareness of the need to diffuse employment and opportunities away from London and south-east England, not least in response to the greatest likely loss of EU labour here, the spatially expressed consequences across the UK of a departure from the EU would be grossly uneven. A UK of exacerbated spatial, structural and social imbalance and inequalities was likely to result. Greatest 'collateral damage' was most likely among those relatively less mobile de-industrialised communities that had voted to leave the EU.

Notes

(1) Although David Cameron fended off demands to hold a straight in-out referendum when he became prime minister in 2010, his then coalition government passed the European Union Act, 2011, which required a referendum for any future EU treaty that granted extra powers to Brussels (Adonis, 2018).

(2) One of Cameron's advisers later admitted (Oliver, 2017) that 16- and 17-year olds should have been given the vote, as had been the case in the September 2014 Scottish independence referendum.

(3) A youth campaign, Our Future Our Choice (Ofoc), was established subsequently to place young people at the forefront of arguing to keep the UK in the EU (Swain, 2018).

(4) This was before the US Trump administration threatened to withdraw from the WTO.

Invited Contribution
3 Tourism and the EU: Retrospect and Prospect

Constantia Anastasiadou

3.1 Introduction

The aim of this chapter is to consider how the EU's approach to tourism may develop or change after the UK has left the Union. Beyond the changes and any exit agreements concluded in terms of trade, aviation and movement of people, capital and services, the UK's withdrawal necessitates reviews of the Union's treaties and of the EU budget. As this chapter argues, some of these reviews and revisions may be relatively straightforward, whereas discussions around the EU's budget are likely to be contentious, generating intense negotiations between member states. In the medium term, policy areas such as cohesion, common visa and consumer protection – all significant for tourism and travel – will change as the existing member states attempt to tackle the aftermath of financial and migration crises and contemplate their future as a bloc.

Tourism has traditionally been cast as a means of promoting European identity and progressing the EU's aims for further European economic and political integration. It is unlikely that the UK's departure will result in a major change to the EU's well-entrenched conceptualisation of tourism as an area of complementary competence. Instead, the existing priorities in terms of employment generation, consumer protection, competitiveness and product innovation are likely to continue but will be conditioned by the broader challenges and changes the bloc experiences as it redefines its longer-term focus. In the medium term, future budgetary decisions and the re-definition of EU policy objectives may impact on the levels of funding made available for tourism-related projects and the types of tourism-related initiatives and actions that are financially supported.

This chapter begins, however, by considering, in retrospect, the pre-*Brexit* relationship between the EU and the UK in terms of tourism, and the EU's existing priorities and actions with regards to tourism. The implications of the revision to the EU common visa policy and the reasons that instigated its change are also considered important in explaining the background against which change is taking place and the EU's responses to these changes. Consideration is also given to broader sociopolitical issues which are shaping the EU, such as shifts in power dynamics that the reallocation of the UK's seats in the European Parliament might instigate, and the European Commission's five scenarios for the future of the EU.

3.2 European Tourism

The tourism sector contributes to over 10% of EU GDP, with the industry directly employing 5% of the total workforce (over 11m jobs) and indirectly employing another 6.6% (in total employing 11.6% of the workforce, over 26 million jobs) (WTTC, 2017). Projections by the United Nations World Tourism Organization (UNWTO) show that international tourist arrivals to the EU28 could reach up to 557m by 2030, up from 380m in 2010, implying an average annual growth rate of around 1.9% (UNWTO, 2017b). These statistics demonstrate that tourism is a major socioeconomic activity in the European Union 'with wide-ranging impact on economic growth, employment and social development and it can be a powerful tool in fighting economic decline and unemployment' (European Commission, 2014: 8).

EU policy making on tourism is a national competence but tourism is mentioned as an area of interest in Article III- 281, in the Treaty establishing a Constitution for Europe. It reads:

(1) The Union shall complement the action of the Member States in the tourism sector, in particular by promoting the competitiveness of Union undertakings in that sector.
   To that end, Union action shall be aimed at:
   (a) encouraging the creation of a favourable environment for the development of undertakings in this sector;
   (b) promoting cooperation between the Member States, particularly by the exchange of good practice.
(2) European laws or framework laws shall establish specific measures to complement actions within the Member States to achieve the objectives referred to in this Article, excluding any harmonisation of the laws and regulations of the Member States. (TFEU, 2012)

This *complementary* approach to tourism is exemplified in the development of the 2007 *Agenda for a Sustainable and Competitive European Tourism* (CEC, 2007), a comprehensive strategy for EU tourism.

According to Panyik and Anastasiadou (2013) the EU has in more recent years adopted a *facilitator* role by focusing on soft 'law' measures, voluntary schemes, partnerships and knowledge exchange. Moreover, EU institutions have created additionality for tourism by focusing on the tourist experience, and have initiated a range of measures to support tourist rights (such as the EC Package Travel Directive) (Anastasiadou, 2008a, 2008b).

In 2010 the European Commission identified four priorities in promoting tourism:

(1) Stimulate competitiveness in the European tourism sector.
(2) Promote the development of sustainable, responsible and high-quality tourism.
(3) Consolidate the image and profile of Europe as a collection of sustainable and high- quality destinations.
(4) Maximise the potential of EU financial policies and instruments for developing tourism. (European Commission, 2010: 7)

Finally, the EU also finances tourism-related activities through grants or indirect funding via financial intermediaries in relation to such EU policy objectives as cohesion or improving European cooperation in research and education (European Commission, 2016). Table 3.1 summarises the most relevant programmes for the tourism sector in the Multiannual Financial Framework (MFF) 2014–2020.

In its review of the future of European tourism in 2014, the European Commission identified the following categories of challenges that

Table 3.1 Most relevant programmes for the tourism sector in the EU

Area	Multiannual Financial Framework 2014–2020
Cohesion	Structural funds: • European Cohesion Fund • European Social Fund • European Regional Development Fund • European Territorial Co-operation
Environment, Agriculture & Marine and Fisheries Policy	• LIFE • European Agriculture Fund for Rural Development • European Maritime and Fisheries Fund
Research, Innovation and Competitiveness	• Horizon 2020 (Framework Programme for Research and Innovation • COSME (Programme for the Competitiveness of Enterprises and SMEs)
Culture and Education	• Creative Europe Programme • Erasmus for All Programme
Employment	• EaSI (EU programme for Employment and Social Innovation) • PROGRESS (Programme for Employment and Social Solidarity) • EURES (European Employment Services) • PROGRESS Microfinance Facility

Source: Adapted from European Commission, 2016.

needed to be addressed by tourism businesses and public stakeholders at regional, national and EU level:

(1) Security and safety issues: environmental, political and social security, safety of food and accommodation and sociocultural sustainability threats.
(2) Economic competitiveness: seasonality, regulatory and administrative burden, taxation, difficulty to find and keep skilled staff.
(3) Technological issues: globalisation of information and quickly advancing changes in technology (digital tools for booking holidays, social media providing advice on tourism services), which pose difficulty in coping with such fast IT developments.
(4) Markets and competition: increasing requests for customised experiences, for new products, growing competition from other EU destinations. (European Commission, 2014: 8)

It can be deduced from this list that current issues for European tourism continue to revolve around sustainability, the capacity to fully take into account 'current and future economic, social and environmental impacts' (European Commission, 2018f), and digitisation, since tourists are increasingly reliant on the flexible and often cheaper digital services of the sharing economy. Understanding how the new business models of the digital age impact on destinations and tourism suppliers is an area where work will continue (Dredge *et al.*, 2016). In 2015 the Commission launched the *Digital Tourism Network*, to discuss how to foster the innovation capacity of tourism entrepreneurs, and is redeveloping the *Tourism Business Portal*, a one-stop-shop, providing information on digital technologies and innovative business practices (European Commission, 2018f).

3.3 Revision of the Common Visa: Implications for Tourism Flows

Currently, citizens from 105 non-EU countries or entities are required to hold a visa when travelling to the Schengen area for short stay visits. A Schengen visa allows travellers to visit the 26 countries of the Schengen area for tourism or business purposes for a maximum duration of 90 days in any 180-day period.

The changed migration situation and increased security threat in recent years have shifted the political debate on the Schengen area in general, and visa policy in particular, towards a reassessment of the balance between migration and security concerns, economic considerations and general external relations (European Commission, 2018a)

The Commission proposed in 2018 'to reform the EU's common visa policy to adapt the rules to evolving security concerns, challenges linked to migration and new opportunities offered by technological developments' (European Commission, 2018c).

The Visa Code has been a core element of the common visa policy: it establishes harmonised procedures and conditions for processing visa applications and issuing visas. It entered into force on 5 April 2010 to facilitate travel and tackle irregular immigration, and to reinforce equal treatment of visa applicants (European Commission, 2018c). The proposed changes to the Visa Code will make it easier for legitimate travellers to obtain a visa to come to Europe, whilst strengthening security and mitigating irregular migration risks (European Commission, 2018i).

Of particular relevance to tourism is the EC's proposed European Travel Information and Authorisation System (ETIAS) which formally entered into force in 2018, but will not become operational before 2021 (Radjenovic, 2018). Under this initiative, visa-exempt third-country nationals are screened against security and migration risks before they start their travel to the EU. The check takes place using digital means and allows visa applicants to apply online (European Commission, 2018b) (see also Chapter 7).

According to the European Commission, the revision of the Visa Code would create several benefits for travellers as it would usher easier, more flexible and faster procedures in applying and processing applications. The scheme would also allow the issue of short-term visas with a maximum duration of seven days in the issuing member state only, to encourage short-term tourism. Travelling would also be easier for frequent visitors who would no longer be expected to apply for a new visa every time they travelled to the EU, but who would benefit from the issue of multiple-entry visas (European Commission, 2018a).

3.4 UK-EU Tourism Relations Pre- and Post-*Brexit*

Access to the single market has ensured hassle-free travel across the EU area which has supported the development of strong tourism flows (WNS, 2018b). As noted in Chapter 1, over three-quarters of international holidays and two-thirds of business visits from the UK are to the EU. In addition, 63% of the UK's leisure travellers come from EU countries, as do 73% of the UK's business travellers. This emphasises how important access to the single market has been for the sector (Pickett, 2016). In 2018 EU residents carried out 25m trips to the UK, which corresponded to 65% of total trips to the UK, whereas UK residents undertook 54m visits to the EU: 75% of the total overseas visits by UK residents (ONS, 2019b; see Chapter 7). Inbound tourists from the EU have contributed around £10bn to the UK economy each year, while outbound tourism to the EU contributes £22.4bn to the UK economy (Tourism Alliance, 2017). These statistics demonstrate that maintaining the existing levels of tourist flows would be mutually beneficial.

As one of Europe's main originating tourist markets, outbound tourism from the UK presents high exposure for such destination countries as Spain, Portugal, Cyprus and Greece (Alogoskoufis, 2016; Garicano, 2016; Taraves & Jorge, 2016). More than 12 million British people visit Spain each year, accounting for around one quarter of all foreign visitors to the country. The decline in value that sterling experienced against the US$ and the euro following the referendum result has made travelling abroad more expensive and led to a reduction of disposable income in the UK and a decline in outbound travel. A further fall in the value of sterling could motivate more UK tourists to holiday at home, which would be encouraging news for domestic tour operators (WNS, 2018a) (see Chapter 7). The southern European states would stand to lose more heavily if sterling continued to remain weak. The combined effects of destination substitution and UK 'staycation' would further undermine the financial position of these European countries (Emerson et al., 2017) who would find themselves disadvantaged because of the UK withdrawal. It is therefore likely that these states would wish to see a continuation of the *status quo* with regards to tourism flows between the UK and the EU.

The UK and EU have indicated that they would like to continue with reciprocal, non-discriminatory, visa-free travel for short-term visits. After the end of the transition period and when it fully comes into force, UK residents will have to apply for an ETIAS visa every three years for a fee of €7. This will allow them to visit EU member states for a period of no more than 90 days within a 180-day period. It is estimated that 95% of the applications will be accepted online. They are also likely to face tougher border controls and restrictions (Whyte, 2018).

3.5 Immediate Issues that Affect EU-UK Tourism Flows

As intimated in the preceding section, freedom of movement has been the main tenet behind the existence of this strong bilateral relationship between the UK and EU (WNS, 2018a, 2018b).

It would be in the interests of both parties to enable a continuation of the *status quo* with regards to tourism. However, this would not be tenable given the UK's decision to leave the customs union and the single market and thus to end freedom of movement after withdrawal. Issues of particular relevance to tourism as part of any future relationship include consumer rights, passenger rights and the package travel directive (Table 3.2) (see also Chapters 6 and 7).

According to Pickett (2016) European air travellers are accustomed to a relatively high level of protection through various directives such as the Package Travel Directive, Passenger Rights, Consumer Rights Directive and compensation and care rights under EC 261/2004, and they would therefore expect to continue enjoying barrier-free access.

Agreements regarding aviation are also important in sustaining visitor flows between the UK and Europe. Intra-regional traveller

Table 3.2 Tourism-related issues needing to be addressed as part of the exit agreement

Tourism-related area of concern	Issues
Aviation	Assuming the UK would leave the Single Aviation Market and the European Aviation Safety Agency, EU law-based rights, obligations and benefits would cease. Ownership control rules and third country restrictions would kick-in.
	Aspects to consider as part of the new agreement included: market access, regulatory matters, level playing field, enforcement and supervision. EU-UK air transport agreement (ATA) and bilateral aviation safety agreement (BASA) to be negotiated.
Freedom of Movement	This is the main tenet of the strong tourism flows between the UK and EU, allowing visa-free travel.
	ETIAS visas will be needed for UK-EU travel after transition. Booking spontaneous European and British holidays could be more difficult for UK and EU citizens respectively.
Passenger Rights	The Air Passenger Rights Regulation No (EC) 261/2004 granting compensation in the events of denied boarding, cancellation and delay is not applicable to passengers of non-EU carriers departing from an airport in a third country. UK passengers could lose £300m a year in compensation in cases of delay if the UK did not retain the EU regulation in some form.
Roaming Charges	The single digital market provides that EU citizens travelling within the EU but outside their country of residence are not charged for roaming services when using their phone while abroad. The law ceased to apply to the UK after *Brexit*, unless there was UK-EU agreement on how these regulations could apply to UK citizens.
Cross-border portability of online content services in the internal market	A new EU regulation would eliminate obstacles to cross-border portability of online content services due to intellectual property issues (geoblocking) to grant 'equal access from abroad to content legally acquired or subscribed to in the country of residence when temporarily present in another Member State such as for holidays, business trips or limited student stays'. EU27 consumers could face geoblocking issues when travelling to the UK after *Brexit*.
Package Travel Directive (PTD) 2015	The UK needed to transpose the requirements of the new PTD 2015 directive into domestic law by January 1, 2018, and bring requirements to force by July 1, 2018. This legislation would remain in place after *Brexit*, unless subject to deregulation. If this situation materialised, the UK and EU would have to reach a reciprocal arrangement that ensured the delivery and use of package travel services remain unchanged and that consumer rights were secured when travelling to and from the EU.
European Health Insurance Card	The European Health Insurance Card may no longer be mutually accepted between the UK and the EU.

Sources: Adapted from: European Commission, 2018e; Kramme, 2017; Pickett, 2016; WNS, 2018a, 2018b.

mobility has benefited considerably from aviation liberalisation policies. In particular, the realisation of the Single Aviation Market and the emergence of low-cost carriers, which now contribute 39% of intra-European seat availability, have encouraged stronger tourism flows (Pickett, 2016).

Determining the future relationship in aviation is likely to be most challenging with the UK's EU exit because 'almost every area of air transport is affected – from access to the internal aviation market and external aviation policy to air traffic management within the Single European Sky' (Vrbaski, 2016: 421). The European Court of Justice (ECJ) is another integral part of EU aviation regulation (Pickett, 2016). The UK government was determined to end the court's jurisdiction, which would, nevertheless, continue through the transition period.

Unless alternative arrangements were put in place, access to the European Common Aviation Area would cease. It would therefore be in the interests of all stakeholders, not least airlines, that a sensible agreement be promoted and even to prioritise air service agreement negotiations between the EU and the UK independent of other negotiations (DG Internal Policies, 2018). The final agreement would have profound implications for the continuation of tourism flows between the UK and the EU, affecting inbound and domestic tourism in both (Chapter 7).

3.6 The Sociopolitical Environment Post-*Brexit* and Potential Impact on Tourism

The effects of *Brexit* on the EU are not only economic and political, but also legal and institutional (Fabbrini, 2016). *Brexit* has created the need to adapt the EU's legal framework to the Union of 27 members. In particular, the remaining EU member states will have to amend the EU treaties such as modifying Article 52 TEU on the territorial scope of EU law as well as change the composition of the European Parliament on account of the UK's withdrawal (Fabbrini, 2016).

Politically, the European Parliament will also change as the UK's seats will be redistributed. It is anticipated that this will be an issue hotly debated between the remaining member states. Historically, the European Parliament had taken a special interest in tourism by providing a platform for civil society and business interests to meet with representatives of the EU institutions and the operation of the Tourism Intergroup, an informal working group comprising MEPs that share an interest in tourism (Anastasiadou, 2008b; Panyik & Anastasiadou, 2013). Post-*Brexit* re-organisation could change existing dynamics with regards to tourism within the European institutions.

More direct impacts on the EU's involvement in tourism are likely to occur in the aftermath of negotiations for the next EU budget which will run from 2021. With UK withdrawal, the EU's system of revenues and expenditures will need to be reviewed and reformed (Fabbrini, 2016). The

UK has contributed in excess of 12% of the EU's gross revenue and, unless net contributor countries are willing to pay more, there will be significant cuts in the next EU Multiannual Financial Framework (MFF) from 2021 onwards (Bachtler & Begg, 2018). For many commentators, the balance in the European Council on economic policy debates would shift post-*Brexit*, with the loss of a large member state supporting liberalisation and budgetary spending controls (Bachtler & Begg, 2018; Irwin, 2015).

An intense period of negotiation will take place between member states to determine whether to sustain current budgetary levels by increasing member states' contribution to the EU budget or to rethink current activity levels to adapt expenditure to a reduced budget. This dilemma is likely to be an issue of contention between the net contributors and the net benefactors to the EU budget. In principle, the EU could reduce expenditures in proportion to the UK quota but this seems unlikely because those member states which are net beneficiaries of EU spending would object to such a decline (Fabbrini, 2016). Alternatively, those member states who are net contributors to the EU budget could increase their contributions to make up for the shortfall in the budget – but it is unlikely they would endorse this approach as they already contribute heavily (Fabbrini, 2016).

The European Commission's (2019) roadmap to an agreement of the EU's 2021–27 budget indicated broad support for streamlining the future budget, radically reducing the number of programmes, while introducing new integrated programmes in areas such as investing in people and the single market and in simplifying funding rules. It was expected that negotiations over the new EU financial framework, scheduled for Autumn 2019, would be highly contentious. In addition, they would be in part conditioned by the UK's final relationship with the EU, and whether the UK contribution to the EU budget would continue in some form in exchange for maintaining access to the single market. Depending on the outcomes of these budget negotiations, the levels and types of funding available for investment in tourism could be significantly affected.

Cohesion policy is one of the most substantial programmes in the EU, with a budget of €352bn over the 2014–2020 period to support less developed regions across the EU. According to Bachtler and Begg (2018) there will be incremental steps to reform this policy area, but there has been a reluctance to undertake more sweeping reforms. Instead, resources may be switched to increase funding for other internal EU policies (such as research, SME development, environment, transport, border security), as well as providing more support for 'external actions', including financing development aid to reduce the flow of migrants from outside the EU.

Proposals put forward for the 2021–27 period recommended an allocation of €330bn along with a reform programme to reduce administrative burden and increase flexibility for beneficiaries of the funds. The number of thematic objectives was also reduced from 11 to 5: Smart Europe, Green Europe, Connected Europe, Social Europe and Citizens' Europe (De Falco, 2019). These changes indicated continued strong support for cohesion policy while adapting it to current EU priorities and vision (European Commission, 2019).

Tourism projects have been funded through the EU's cohesion policy and the European Structural and Investment Funds (as indicated earlier in Table 3.1), so reform of the policy will have implications for the types of actions and projects supported from 2021. Further, the European Parliament (2019) requested the addition of a separate budget line for tourism in the next multi-annual financial framework (MFF), a significant development acknowledging the importance of tourism as a socioeconomic activity. And the Council of the European Union (2019) endorsed continuation of the European Commission's work in supporting the sustainable development of tourism within the EU.

3.7 Issues for the EU and European Tourism: A Longer-term View

In considering the long-term approach to tourism, it is unlikely that the UK's departure will generate a major change to the EU's tourism competence. Evaluations of current practices in tourism indicate a preference for the continuation of the current path of complementary competence and existing types of policy instrument interventions (collaboration, knowledge exchange, voluntary schemes). For example, Weston et al. (2014) argued that Europe will face increased competition from other regions and suggested that the EU continued to promote itself as a leading heritage and cultural destination. The authors also identified four areas where the EU could further support tourism development: developing quality sustainable tourism; supporting the development of SMEs; harmonising the accommodation classification systems; and promoting the development of 'slow travel' (Weston et al., 2014: 9). These are barely different from previous conceptualisations of the EU's approach to tourism (see Panyik & Anastasiadou, 2013, for a more detailed discussion).

The European Commission carried out a public consultation on the future of European tourism to better identify key present and future challenges and opportunities and to help revise, if necessary, the Action Plan for the sector put forward by the European Commission in 2010 (European Commission, 2014). The outcomes of the consultation for the Future of EU tourism also indicated that changes would be gradual. There would be no major overhaul of the existing approach as themes had tended to be repeated in previous similar exercises: preference for EU intervention on product development and promotion together with investment in infrastructures, networking and training but avoiding harmonisation of standards and increased regulation for the sector (European Commission,

2014). These outcomes indicated a desire to focus EU intervention and support in particular areas of tourism production.

One area that is likely to have a growing significance is the need to address the impacts of new technologies, digitalisation and the implications of the shared economy for destinations as well as for the employment rights of workers in tourism and hospitality sectors. It is anticipated that specific actions or programmes may be developed on these topics in response to the gaps and differences in existing regulatory environments within member states (Dredge *et al.*, 2016).

A wider issue that will underpin any future discussions about tourism will be the shape the EU will take after *Brexit*. Predictions made immediately after the referendum vote suggested that *Brexit* would encourage further Eurosceptic tendencies and precipitate requests for similar referenda in other European countries. However, a number of researchers (Collins, 2017; de Vries, 2017; Fabbrini, 2016) claimed that the opposite was true. De Vries' (2017) research found that far from encouraging the rise of Eurosceptic movements in Europe, since the UK's decision to exit, the EU populist and anti-EU parties had been declining in popularity, and support for EU membership had actually increased. She asserted that as the uncertainty of leaving manifested itself, an increase in support for EU membership was especially pronounced among those who thought the UK would fare badly after *Brexit* (de Vries, 2017). Collins (2017) also concluded that *Brexit* appeared to have galvanised Europe behind the concept of European unity. Subsequent election results in Italy and elsewhere might suggest otherwise.

Europe has to deal with challenges as diverse as migration, climate change, terrorism and the transition to a digital economy, as well as the need to boost growth, jobs and investment (Bachtler & Begg, 2018). For some commentators, losing Britain's 'Anglo-Saxon' economic influence with its strong support for free trade could lead to an inward looking, protectionist EU (Oliver, 2016).

Oliver (2016) identified three possible scenarios. The first sees the UK's departure upsetting complex relations within the EU between north and south, east and west, small and large members, liberal free-trading states and ones more inclined to protectionism. Franco-German relations, often considered the motor of European integration, have often seen the two countries using the UK to counter-balance each other. The EU's institutions would lose their British influences, with the EU having to renegotiate its qualified majority voting (QMV) system,[1] national quotas and to make up for the loss of the UK's budget contribution. The outcome could be a more confused, divided and weakened EU (Oliver, 2016).

The second scenario would see Germany's position in the EU strengthened. This would enhance German support for further integration, and strengthen the Eurozone and EU institutions (Oliver, 2016).

The third scenario would see Germany remain ambivalent about leading or further developing the EU, with other states such as France also being wary of any growing power in Brussels.

The European Commission itself (2017c) identified several drivers of change for the next decade:

- an ageing population;
- a decline in the EU's share of global GDP;
- the legacy of the financial crisis of long-term unemployment and high levels of public and private debt;
- heightened security threats; and
- a mistrust of EU institutions and a questioning of their authority and legitimacy.

Against these challenges the European Commission (2017c) proposed five scenarios of how Europe could evolve by 2025 to direct the way these issues might affect its economies and societies: carrying on; nothing but the Single Market; those who want more do more; doing less more efficiently; doing much more together. These scenarios were proposed as starting points to encourage an open conversation amongst the EU27 on how to move forward as a Union.

The European leaders' main objective has been to maintain unity (Anghel *et al.*, 2018). The Bratislava declaration and roadmap (2016) set out an agreement of how the EU should move forward post-*Brexit*: namely to have separate discussions on the future of Europe and on the EU's policy priorities, often within the framework of Leaders' Meetings. Subsequent further pledges were made: in Rome (2017) to boost the legitimacy of the EU, connect with disaffected citizens, and to combat Euroscepticism (Anghel *et al.*, 2018); and in Sibiu (2019) to stay together and embrace the future as a new Union of 27 (European Commission, 2019).

3.8 Conclusion

The aim of this chapter was to consider, in retrospect and prospect, how the EU may review its approach to tourism subsequent to UK withdrawal. To this end, the chapter considered both direct and indirect areas impacting on tourism: legislation that influences tourism mobility and the availability of funding to support tourism innovation; and how changes to the institutional, legislative and broader socioeconomic environment precipitated by *Brexit* may affect the EU's approach to tourism.

The analysis demonstrated that the tourism flows between the UK and the EU have flourished within an environment that supported the freedom of movement. Considerable uncertainty persisted in withdrawal negotiations concerning how critical issues relating to consumer protection, visa policy and aviation would be determined (see also Chapters 6 and 7). These decisions would have a significant impact on intra-European tourism flows and the shape of the tourism and travel sector after UK withdrawal. It would be in the interests of industry and consumer groups for a continuation of the status quo. Other EU member states, whose tourism-related sectors would also be profoundly affected, will have taken a very close interest in the outcomes of the negotiation process.

In the short term, significant change is unlikely for tourism through the regulatory and treaty changes that the UK departure necessitates, because an emphasis will be placed on streamlining and simplifying existing regulations for the remaining EU27.

An overview of recent consultations and reflections has illustrated that in the medium-term, at least, within the EU there will be a continuation of the current approach to tourism with only marginal reviews to consider the impact of new technologies on tourism destinations. In terms of policy approaches, there will be continuity of action in the areas of heritage and cultural tourism, digitisation, the impacts of new business models and disruptive economy on tourism destinations. There may, however, be changes to policy intervention instruments as a result of institutional and budgetary reforms.

Tourism will thus continue to be considered important for the European integration process, and the emphasis that has been placed on the promotion of cultural heritage as an area of shared interest is likely to continue. The need to reinforce notions of unity, common heritage and shared identities may bolster further the discourses that support tourism's input and contribution to the creation of a common European identity.

Note

(1) QMV is the means through which decisions are reached on certain policy areas.

4 Theorising *Brexit* and Tourism

> *It's very Balkanic what is happening in Britain. Deal, no deal, soft border, hard border, no agreement. It's the Balkans! ... While we are trying to Europeanise, it looks like they are Balkanising.*
>
> Edi Rama, Prime Minister of Albania,
> (quoted in Walker & Triest, 2019: 29)

This chapter explores a number of theoretical approaches that may hold potential for examining the *Brexit*-tourism nexus.

4.1 *Brexit* as an Object of Study

> Until the Brexit vote happened, contemplating Brexit or the withdrawal of any member state from the EU was something of a taboo topic for many ... It means there has been a scramble to understand and analyse such topics as European disintegration. That leaves us with a lack of relevant and rigorously researched literature. A lot ... will be conjecture. Due to the polarising nature of Brexit, for both students and teachers the task of being able to critically think and analyse this literature will be an important challenge for avoiding the inherent bias in many people's work. (Oliver & Boyle, 2017: 3)

As noted in Chapter 2, a range of arguments was advanced for explaining the referendum outcome: that it was the largest popular rebellion against the establishment in the UK's modern history, a backwash from those groups and sections of the population most affected (and disaffected) by the impacts of austerity politics (Wahl, 2017; Worth, 2017), or an expression of a globally resurgent ethno-nationalism challenging key aspects of neoliberal globalisation from the political right (Galbraith, 2017; Patomäki, 2017). Hobolt (2016: 1259) expressed the outcome as a divide between the winners and losers of globalisation; Toly (2017: 142) as economic inequality, political disenfranchisement and social exclusion at the regional level driving a fresh interrogation of the relatively open world order; and Pettifor (2017: 127) as those 'left behind' protecting themselves from the predatory nature of market fundamentalism (a notion challenged by Antonucci *et al.*, 2017).

For others, the referendum vote created a discontinuity whereby previous norms and rules of engagement no longer automatically applied, with earlier accepted values and practices exposed to renegotiation (Guldi, 2017). It brought into question both the unity of the United Kingdom and the continuation of the European Union itself (Bristow *et al.*, 2016), embodying 'emergent and contested political projects of belonging' (Cassidy *et al.*, 2018: 188).

Goodhart (2017) contended that the (pre-existing) cleavage in British society exposed by the referendum vote represented two large value clusters. One was of the educated and mobile who see the world from 'anywhere' and who value autonomy and fluidity. By contrast, the more rooted, generally less well-educated who see the world from 'somewhere', prioritise group attachments and security and feel/are threatened by the local impacts of global processes such as de-industrialisation and immigration.

As well as stirring up the murky waters of racism, regional, class and generational division (e.g. Kelly, 2016), the referendum vote also appeared to have bred 'a sinister strain of anti-intellectualism' (Wright, 2016). The role of experts and intellectuals was spurned and ridiculed by some politicians, most notoriously, when as part of the 'Leave' campaign, Michael Gove was reported as refusing (being unable) to name any economists supporting *Brexit*, on the spurious grounds that 'people in this country have had enough of experts' (Mance, 2016). Further, some proponents in the referendum campaign and beyond appeared to take a particularly cavalier approach to facts, political promises and the meaning of 'objective truth', exacerbating the distrust in democratic institutions and elites that appeared to encourage a negative referendum vote in the first place (see Becker *et al.*, 2017).

Questions over the legitimacy and relevance of experts and the nature of their knowledge therefore rendered the wider nature and process of the UK's withdrawal from the EU a core concern for academics, not least in terms of reflecting on our own knowledge, practice, impact and relevance (Bristow *et al.*, 2016), and with manifold consequences for UK higher education (see Chapters 6 and 13).

4.2 *Brexit*, Tourism and the Intimacy of Geopolitics

The act of the very first country seeking to leave the European Union is an overtly geopolitical proposition, albeit one deriving largely from domestic political considerations. As such, the nature and impacts of *Brexit*, not least those pertaining to tourism and related mobilities, may be viewed and interrogated from geopolitical perspectives.

While tourism is just one of a number of actors that may influence developmental processes at local, regional, national and supranational levels, so tourism in its turn may be impacted upon by both deliberate policy and happenstance from across a wide range of political, economic, cultural, environmental and conflict realms. Tourism policy *per se* may be little more than a blunt instrument in the face of multiple exogenous agencies. In the case of the EU, for example, as an area for 'national competence' little in the way of explicit tourism policy has been promulgated (Chapters 1 and 3). Yet tourism processes have been deeply influenced by policy decisions in areas which impinge upon the nature and practice of tourism, such as the freedom of movement of people, goods and ideas, structural fund policy, and biosecurity.

As such, tourism is deeply embedded in politics (and, indeed, politics in tourism) at all levels, contributing significantly to global ordering (Franklin, 2004; Tribe, 2008), while embracing both the symbolism and reality of 'neoliberal hegemony' (Monbiot, 2016a; Sayer, 2014).

(International) tourism is an implicitly geopolitical activity in its reliance upon working relationships between blocs', countries' and regions' administrations and businesses (e.g. Macleod & Carrier, 2010; Weaver, 2010). And complementing such tourism-related mobility is the need to facilitate cross-border flows of the capital, commodities, labour and skills required to sustain the (international) tourism process (Hannam, 2013; Reiser, 2003).

The spatial consequences of international relations at supra-national and national levels can impact substantially on activity at sub-national levels, both at tourism destinations and at international borders. But local tourism-related responses to such impacts can vary considerably, are often place- and context-related, and thereby articulate variable local conditions (e.g. Dołzbłasz, 2017; Laine, 2017). Hazbun (2004), amongst others, has referred to this as re-territorialisation: the increased relevance of location and the particular characteristics of place. In the context of the UK's withdrawal from the EU, the particular geopolitical considerations surrounding the nature of the UK-Irish border (Chapter 9) and the position of Gibraltar (Chapter 11) are striking examples of particular circumstances of place and time confounding broader geopolitical assumptions.

More than 2 billion international tourism movements are predicted to take place annually by 2030 (UNWTO, 2014). Tourism has become an inextricable part of neoliberal hegemony driving globalisation processes: a two-edged sword. On the one hand, international tourism may be seen to symbolise the mobility ideals of the European Union, but on the other, a significant element of the UK vote to leave the EU appeared to be directed against such concomitants of globalisation as labour migration, foreign second homes and faceless corporatism. Is *Brexit* therefore implicitly anti-tourism, expressing rejection of the ideals of culture contact? Or is such a paradox explicable because the association of tourism with culture contact (or at least cultural engagement) may not be a foremost objective for many UK tourists (see Kock *et al.*, 2019)?

As the 2016 referendum vote was not expected to return a majority in favour of leaving the EU, it was evident that little in the way of strategy existed either to steer the UK through a painful negotiating process or for articulating the UK's international position after *Brexit*. That the hastily contrived notion of a 'Global Britain' implicitly entailed the movement of people, goods and ideas, and thus encouraged an enhanced international tourism environment, would seem to contradict both the above hypothesis and the 'Take back control' *mantra* of 'Leave' proponents. A swiftly concocted post-referendum *Tourism Action Plan* document only acted to emphasise the policy vacuum that tourism and a post-*Brexit* UK faced (Chapter 5).

Since the 'cultural turn' in the social sciences, a growth of critical studies in tourism and of critical geopolitics has brought new methodological insights into areas of research that had previously remained largely positivist and relatively isolated. 'Intimacy geopolitics' is one such area of study emerging from feminist engagement (Pain & Staeheli, 2014), offering considerable potential for the critical study of tourism (e.g. Dowler, 2013). Intimacy geopolitics arises from the view that intimacy is wrapped up in national and global geopolitical processes and that geopolitics is created by, and consists of, relations and practices of intimacy (Pratt & Rosner, 2012). Although intimacy geopolitics has been criticised for negatively focusing on violent oppression (physical, emotional and psychological) as a means of exerting control, a positive strand of intimacy geopolitics embracing hospitality and diplomacy has emerged to offer potential for tourism-related study.

An example is Ruth Craggs' (2014) work on aspects of Commonwealth diplomacy, where she has argued that hospitality is an important part of geopolitical practice. She points to ways in which international relations can be highlighted as performance through hospitality, a conceptualisation that 'makes space for diplomatic labour, the construction of atmosphere, and the often uneven power relations that such performances embody' (Craggs, 2014: 90).

In the US, 'gastrodiplomacy' refers to 'the confluence of public diplomacy and national cuisine' and is concerned with 'the meal as a symbolic and real tool in the acquisition of soft power among countries' (Moscato, 2018: 187). As a variation of public diplomacy it enjoys a longstanding political tradition in the US. In 2012 for example, Secretary of State Hillary Clinton launched a Diplomatic Culinary Partnership programme for the White House, promising visiting dignitaries 'nuance and nutrition' and a vision of enhanced diplomatic relations (Moscato, 2018: 187). If soft power is about shared values (Sun, 2012) (see below section 4.7) and 'cultural congruency' (Entman, 2008), the symbolism of a shared meal as a form of understanding between nations can help project a narrative about where the nations are actually reaching understanding (Moscato, 2018: 194).

These two strands of complementary literatures would suggest a fruitful area of investigation into the hospitality performances undertaken during the UK-EU exit negotiations. In addition to more formalised, public and therefore politicised interactions, how far have the performances of UK politicians, officials and diplomats in informal, 'hospitality' contexts with their EU/EC counterparts (and, perhaps, among themselves), influenced the course of geopolitical events?

4.3 *Brexit* as Transformation

4.3.1 Reversed path dependency

The UK majority vote to leave the EU acted to break the pattern of path dependency that the 'deepening and widening' aims of the Union had established for its member states (Chapter 1). Interconnecting strands of social, economic, cultural and environmental policies and political strategies presented a formidable web of relationships and commitments that was grossly underestimated by *Brexit* advocates and/ or consciously targeted for explicit disentanglement.

Path dependency reversed? Such a conception would complement the empirical observations of Lim (2018) and others in their evaluation of processes relating to a country's withdrawal from a supranational union. The most obvious 'gain' such a union may have brought to tourism development – greater and seamless accessibility across national borders, facilitating increased tourist numbers and receipts – is unravelled by *Brexit*, as may be the social and economic benefits that have accrued (Wong *et al.*, 2011a, 2011b). Thus again, within a European context, *Brexit* might be viewed as implicitly anti-tourism.

4.3.2 'Hesitant transition'

The concept of 'hesitant transition', although rarely defined, has been employed to qualify aspects of 'lagging' socioeconomic 'transition' (largely in relation to Central and Eastern Europe), in contrast to the immediate and substantial impacts of 'shock therapy' policies (e.g. Gentile & Marcińczak, 2014). In relation to processes involved in the UK's leaving the EU, 'hesitant transition' can perhaps be applied in two senses.

First, it can be observed relating to the hesitant negotiations towards exiting undertaken by the UK government. Extended over more than two years, these were characterised by political division within the ruling British Conservative Party, and the persisting mismatch, noted earlier (Chapter 2), between the ambition of leaving the EU and a coherent understanding of the tangible objectives of such an action, resulting in uncertainty and further hesitance (for example in capital investment decisions) across a wide range of business sectors.

Second, application of the notion of 'hesitant transition' recognises the different trajectories of different sectors – a 'variety of paths of socio-spatial development' (Marcińczak et al., 2014: 1399). This also reflects the concept of 'multiple transformations', which characterises different spheres of socioeconomic endeavour being transformed at different paces (Sýkora & Bouzarovski, 2012).

Exemplification can perhaps be seen in the differential impacts upon various UK sectors of both the referendum vote to leave the EU and the actual process of exiting. Sub-sectors within tourism (attractions, accommodation, hospitality, events), for example, will experience impacts within different time frames, with varying speeds and in relation to different aspects of the dynamics of EU withdrawal.

Such concepts would appear to offer analytical potential for closely examining impacts of *Brexit* on various tourism and hospitality sub-sectors.

4.4 *Brexit*, Tourism and 'Crisis'

While many crises may be sudden and unforeseen ('natural' disasters, terrorist attacks), and others may be anticipated but without a specific time frame (political crises, international conflicts and tensions), the UK referendum on whether to remain or leave the European Union was held on a specific date which had been known for some time, and which could have resulted in one of only two possible outcomes. And yet, because the outcome was largely unanticipated (the referendum would not have been held otherwise), it resulted in an immediate crisis of European capitalism, or perhaps more accurately, a series of crises, sequential and cross-cutting, including a sudden fall in the value of sterling and threats of FDI retrenchment in the UK.

Hall's (2010) review of the literature on tourism and crisis suggested that economic and financial dimensions received the most research attention, albeit often linked to such other elements as terrorism-related incidents or rising energy costs. Hall's review challenged the way crisis was conceptualised in relation to tourism, and questioned what actually constituted normality (as opposed to a crisis).

Strategy reorientation, tourist market supply adjustments (Cocean & David, 2011), and restoration of consumer confidence (Lim, 2017) have featured prominently in the debated role of tourism in the physical and psychological reconstruction of post-crisis situations (e.g. Avraham, 2016; Nurković & Hall, 2017).

A crisis adversely affecting one tourist destination may benefit others (Cirer-Costa, 2017), and under such circumstances the appeal of traditional destinations can strengthen; while large tourism companies can better adapt to unexpected circumstances than can smaller ones. But Cró and Martins (2017: 5) questioned the definition of a *tourism* crisis. 'Crisis' can be a complex and chaotic (almost by definition) phenomenon (Brecher, 1978), and recovery of the tourism sector from a crisis may be far more complicated than for other sectors. Strong partnerships and coordinated action would appear essential for successful recovery (Cavlek, 2002: 487; Cró and Martins, 2017: 5), with each stakeholder's participation required to secure such critical objectives as: successfully rebuilding destination image, overcoming adverse policy impacts from other sectors, effective management of media coverage, reduction of mobility constraints, and business and consumer regulation support (Steiner *et al.*, 2013).

In relation to the UK referendum vote to leave the EU, two major crisis dimensions could be observed: (i) the contextual and (ii) the consequential; the referendum vote being born of, and helping to create, crises.

4.4.1 The contextual crisis/crises

As a longer-term context for the referendum vote, there existed, if only acknowledged in hindsight, pent-up, festering frustration at the continuing negative consequences of the 2007+ 'financial crisis'. 'Austerity' measures had 'left behind' UK regions and localities with crumbling infrastructures and diminishing community support. Voters of such areas were led to believe that, by some unrealistic sleight of hand, in voting to leave the EU they could 'Take back control' (presumably from the EU/EC's 'faceless bureaucrats') of their own lives and communities. This fed on one of the unfolding effects of the 'financial crisis' that had induced a restructuring and reorganisation of consumption, production and administration (Visser & Ferreira, 2013: 4).

Studies addressing tourism responses to earlier aspects of the 'financial crisis' in Europe (e.g. Papatheodorou *et al.*, 2010; Smeral, 2009, 2010) confirmed, unsurprisingly, that the effects were often far from evenly felt through society. In the UK, Eccles (2011) found that from a survey sample of 2000 people, over a quarter intended to forego their holidays completely. Others searched for better value late deals on the internet. A growing preference for UK 'staycations' had been predicted to particularly benefit Scotland (Coles, 2013: 40).

Reinhardt's (2011) research in Germany also revealed a higher proportion of trips being undertaken within (rather than outside) the country. These were generally shorter in duration with less money being spent on them: the intensification of an ongoing trend. However, reflecting socioeconomic differences, between 2004 and 2010, blue collar workers (41%) were half as likely to travel as civil servants (80%).

4.4.2 The consequential crisis/crises

Stability – political, economic, cultural – is an obvious prerequisite for tourism development. Both at borders and tourism destinations, the interactions between stakeholders constitute fundamental dimensions of connectivity. Such connectivity – the mobilities and interactions associated with tourism-related processes – can be swiftly subverted by political, perhaps geopolitical, acts and declarations. The uncertainty that the *Brexit* vote and the subsequent long period of slow negotiation brought, seriously threatened such stability for the UK travel and tourism sector as well as for related mobilities and their economic impacts in, for example, both Northern Ireland and the Irish Republic (Chapter 9).

Further, regional tourism organisations may be weak and vulnerable in the face of the power of higher level (geo)political decisions and the response behaviour of transnational tour companies, accommodation providers and airlines. Faced with a post-referendum fall in share prices, the ability of such transnational tourism-sector providers to rapidly transfer their business from one destination – region or country – to another (e.g. Farmaki, 2015) in a perceived crisis, mimics the behaviour of transnational manufacturing corporations.

Both the UK's and the EU's tourism-related dynamics would eventually evolve from *Brexit*-related crises. In placing these in perspective, the social systems theories of Luhmann (1982) are instructive in viewing society as a complex structure consisting of many independent sub-systems capable of adapting to 'turbulent environments' by pursuing internal change. Thus, for example, from a tourism-sector perspective, Janson (2017) pointed to four major areas in which UK policy would need to undertake internal adjustment in order to be able to evolve away from a post-*Brexit* crisis environment: amending the Package Travel Regulations (Chapter 7), compensating for the loss of the Common Agricultural Policy (Chapters 3 and 6), developing a new aviation strategy (Chapter 6), and revising border control requirements (Chapters 3, 7).

If this volume had had the luxury of a longer gestation period beyond transition and into the promised glorious 'Global Britain' future, resilience would have been invoked as a central concept for examining medium- to longer-term institutional and destination responses to the consequential crises of the UK's EU withdrawal. The role of human capital and notably 'a skilled labour force that enhances rapid process and organizational innovation' (Crescenzi *et al.*, 2016: 28) has been highlighted as important in helping to strengthen local and regional resilience. The UK government acknowledged this 'higher skills' requirement in its proposed post-*Brexit* immigration policy (Chapter 6), but possibly at the expense of the critical need for lesser skilled labour in such sectors as hospitality and social care.

Having emerged as a popular academic tool 'seeking to capture the differential and uneven ability of places to react, respond and cope with uncertain, volatile and rapid change' (Pike *et al.*, 2010: 1), local and regional resilience will be a closely researched area in the months to come.

4.5 Identity Through Peripherality/Detachment?

One of the constant themes running through, or at least underlining, the UK's relationship with 'Europe' (Chapter 1) is the islands' physical and psychological peripherality. In terms of European integration, this had been symbolised by the UK's opting out of the single currency and not entering the Schengen agreement. The arguments for British 'difference' have usually entailed citing the absence of land borders (until the 1920s at least), not being invaded or occupied since 1066 (the German occupation of the Channel Islands during World War II notwithstanding), and not being 'defeated' in Europe (yet *England*'s footballing achievement of 1966 needs to be forever resurrected as some psychological reassurance of ... what exactly?).

Northern Ireland is doubly peripheral, with the added resonance of UK governments appearing not to care about or understand the province, except when wider security has been threatened or voting support required.[1]

Prior to Scottish and Welsh devolution in 1999, Lowenthal (1994: 15) could argue that British centrism ignored or belittled outlying regions, and that 'national' was normatively *English* (now articulated, for example, in the apparent (con?)fusion of *VisitEngland* and *VisitBritain*).

Following the Maastricht Treaty (1992), the UK was the only member state of the European Community not planning to (translate and) publish 'the composite anti-nationalistic' new *Histoire de l'Europe*, on the grounds that, in the words of the Eurosceptic Teddy Taylor[2] 'this country's schools are traditionally used in the dissemination of knowledge and not for propaganda' (cited in Lowenthal, 1994: 16).

One literary historian has claimed for *England* the world's 'most strongly defined sense of national identity' (Ousby, 1990: 2). Lowenthal (1994: 20) saw this as having largely derived from the nature, perception and representation of *English* landscape. 'Nowhere else is landscape so freighted as legacy': it is the *Leitmotiv* (the ironic use of a 'foreign' word apparently passing without comment) 'of the "solid breakfasts and gloomy Sundays" of Orwell's' (1941: 193) *England*. And perhaps it is the fact that apart from a relatively small tract of Caledonian forest, the British landscape has been moulded and modified by human endeavour (and mistakes) like no other, and is consequently more closely associated with national (self-)imagery. But for *Brexiters* the imagery loops back to an untenable imagined past (O'Rourke, 2019): the conceit of a supposed shared national self-identity that appeared to sustain the 'Leave' trope of 'Taking back control' (Chapter 2).

Although contested identities are usually associated with cultural difference (Nicholson & Marquis, 2015), the identity cleavages between 'Britishness' and 'Europeanness' represented by the referendum vote appeared to reflect geographical, socioeconomic (Hobolt, 2016) and age-related (Matti & Zhou, 2017) factors. The 'betwixt-and-between'-ness (Bauman, 2004: 29) of the ambiguous search for identity within 'liquid modernity' (Bauman, 2000) was well exemplified in the multi-layered reasonings behind voters choosing to leave the European Union as well as the often misleading and sometimes patently mendacious arguments employed to induce such a vote (see section 4.1 above).

The UK's insularity has acted to reinforce the mythical sense of (largely *English*) British identity. While outside observers from Virgil to Gore Vidal have viewed such insularity as 'marooning' Britain from the rest of the world, those politicians who have sought to reverse UK membership of 'Europe' have found satisfaction, even justification, in physical isolation.

> Our Continental neighbours use 'insular' as a term of abuse, but we [sic] in Britain have every reason to be thankful for our insularity. Our boundaries (that troublesome [sic] one in Ireland apart) are drawn by the sea … Unlike those of most other nations they have not been drawn, rubbed out and redrawn time and time again … The blessing of insularity has long protected us against rabid dogs and dictators alike. (Tebbit, 1990, cited in Lowenthal, 1994: 22)

The language has been updated, but the emotional sense of divisive insular superiority amongst the leading advocates of the UK's EU withdrawal has remained much the same. But how can such an attitude be justified or even contemplated by any balanced person in today's hyper-connected world? Who could believe that this same UK is currently the second largest source of international tourists in Europe?

4.6 The UK Passport: Identity Symbol for *Brexit*?

> *Her Britannic Majesty's Secretary of State Requests and requires in the Name of Her Majesty all those whom it may concern to allow the bearer to pass freely without let or hindrance, and to afford the bearer such assistance and protection as may be necessary.*

The above imperial requirement, as unrealistic as it may now be, still adorns the inside front cover of the UK passport. This document confirms an individual's ability to travel away from and back to a country that does not otherwise require its citizens to carry identity cards. It symbolises, rather like the motor car, a putative, mythical freedom to 'go anywhere, do anything'. Unlike the motor car, the passport cannot be customised, but its physical nature and the message it conveys can stimulate strong feelings of identity, and, perhaps, of certain rights.

Thus, the UK passport encapsulates concepts of individual and collective identity, travel behaviour and the spatiality of border crossing. It also carries evidence of other states' ability to screen, manage and control access to their territory (Neumayer, 2006).

When, in December 2017, the UK Home Office announced that British passports would once again have 'blue' covers following EU withdrawal in 2019, it was, for some, a symbolic encapsulation of the emotions and rhetoric surrounding *Brexit* (Table 4.1), and its associated *mantra* of 'Taking back control'.

'Blue' (actually black, or at best blue-black) was first used for the cover of the UK passport (Figure 4.1) in 1921. The design was changed in 1988, some 15 years after the UK had joined the then EEC, to the smaller burgundy 'EU' design (Figure 4.2).

Table 4.1 Emotional/rhetorical values attached to the 'blue' UK passport

Speaker	Emotion/rhetoric
Theresa May, as UK Prime Minister	The UK passport is an expression of our independence and sovereignty – symbolising our citizenship of a proud, great nation. That's why we have announced that the iconic #bluepassport will return after we leave the European Union in 2019
Nicola Sturgeon, Scotland First Minister	… insular, inward-looking, nonsense
Brandon Lewis, Home Office minister	Leaving the EU gives us a unique opportunity to restore our national identity and forge a new path for ourselves in the world … One of the most iconic things about being British is having a British passport
Ed Miliband, former leader of the Labour Party	It is an expression of how mendacious, absurd and parochial we look to the world
Nigel Farage, former leader of the United Kingdom Independence Party	You can't be a nation unless you have this symbol
Mitch Benn, columnist and comedian	The whole 'return of the blue passports' nonsense is in many ways the perfect symbol for Brexit, in that it'll cost a fortune, it'll achieve precisely nothing apart from making some idiots briefly happy for entirely invalid reasons … it's based on a completely spurious notion of a non-existent glorious Britannic Golden Age
Anon, *The Guardian* editorial	Any national identity imperilled by the colour of its documents must be pretty feeble to begin with
Prof Steve Peers, Essex University	The reality is that the new passports will symbolise having fewer free movement rights … It does seem odd to make a big patriotic noise about something that makes it harder for you to travel
Guy Verhofstadt, European Parliament chief *Brexit* coordinator	… the UK could have had any passport colour it wanted and stay in the EU.

Sources: Anon, 2017a; Benn, 2018; Cordon & Wardle, 2017; Roberts & Rankin, 2017; Townsend, 2017.

Figure 4.1 The author's last 'blue' passport: issued in 1988, the last year for its type, with fine copper plate handwriting courtesy of the Peterborough passport office. While *Brexiters* have obsessed over the colour, perhaps it was rather the larger size and nature of the stiffened board covers of the document that offered, for some, a more reassuring sense of resolve and longevity

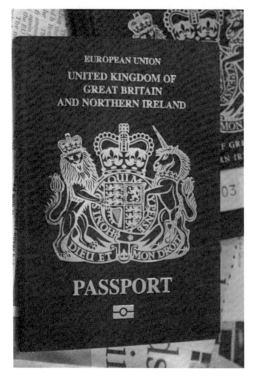

Figure 4.2 The 'EU' burgundy passport

Cutting short chauvinistic *Brexitacious* fervour in the wake of the December 2017 Home Office announcement, Brussels and other sources pointed out that under a system first agreed by the Thatcher government in 1981, the UK had never been legally obliged to use the same burgundy passport design as most other EU member states, but had agreed to do so in a joint resolution of member states in the European Council. There was an opt-out clause which the UK government had chosen not to take up, although it has been employed by some states, most recently Croatia, which has issued blue passports since its accession in 2013.

Most changes to passports since the 1980s have been mandated by international aviation and security agreements framed outside of the EU, and these will continue to dictate the size and content of the travel document (Roberts & Rankin, 2017).

Thus 'Blue' was not, after all, a legitimate part of the atavistic *Brexit* iconography. And the tendering process, required under EU procurement rules, for printing the post-*Brexit* 'Blue' passport, as if the symbolism could not become any more obvious, was inevitably lost by the UK's De La Rue to the Franco-Dutch company Gemalto. *The Daily Mail* claimed a resulting 'wave of fury' washing through the country. The government counter-claimed that about £120m would be saved over the five-year contract, while 70 jobs would be created at some of Gemalto's five UK plants (Walker, 2018).

Inevitably, the 'Blue' passport was not ready in time to be issued for 30 March 2019, but UK passports issued from that date were no longer to include the words 'European Union' (UK Government, 2018d). This requirement would need to be postponed by seven months to coincide with the revised withdrawal date of 31 October 2019.

4.7 Reputation, Re-branding and Soft Power

4.7.1 National re-branding?

The twin *Brexit* mantras of 'Taking back control' and 'Global Britain' offered a crude example of attempted national re-branding which, in the first case, appeared to be both responding to the medium-term crisis of financial turmoil since 2007 and to an immigration 'crisis'. The second repeated slogan anticipated the existential (*exit*stential?) crisis that was likely to follow the UK's withdrawal from the EU (section 4.4 above).

There can be a number of reasons why a country might want to convey a better image of itself (Kiambi & Shafer, 2018: 175). These include being part of a process to assist recovery from a crisis, whether it be natural disaster, war or economic depression (Insch & Avraham, 2014), or to encourage investment and tourism,[3] not least for a country seeking (wider) recognition after exiting a supranational union.

Thus, following the 2016 referendum result, the UK government's subsequent tourism (DCMS, 2016) (Chapter 5), industry (UK Government, 2017a; Prime Minister's Office, 2017; Kent, 2017) (Chapter 5) and environment (DEFRA, 2018b; UK Government, 2018a) (Chapter 8) strategies for a 'post-*Brexit* Britain' could be seen to be components of a rebranding exercise for the UK national (self-) image, a propaganda trajectory aimed at both domestic and international audiences, motivated by soft power geopolitics and economic necessity.

In holding far greater importance than simply as tools for attracting tourism and investment, such nation branding campaigns can contribute to the articulation and meaning of national 'independence', and, in the case of exiting the EU, can contribute to a re-conceptualisation of the local and regional geographies of Europe (Kaneva & Popescu, 2011: 203). International tourism can contribute to such a rebranding exercise through its potential attribute of 'soft power': the ability to influence others (to achieve an objective) by attraction and/or persuasion, rather than by coercion or payment (Nye, 2008).

Tourism can thus become a propaganda tool: its symbols and arte-facts accruing meaning and implication, while its stakeholders become actors on the stage of image projection and 'public diplomacy' (Davis Cross & Melissen, 2013). While this key component of soft power geopolitics has become increasingly enacted in virtual space, away from the altogether less presentable realities of destinations' quotidian, its consequences are tangible in terms of international mobilities.

Major cultural events, such as European Capital of Culture conferment (see below), can generate considerable 'visitability' and provide the context for both (international) prestige and (domestic) reflection, including role and image reappraisal.

Although tourism's unpredictability can render soft power less effi-cacious than proponents might suggest (Fan, 2008), it would seem to have a dual role to play here. On the one hand, a country's outbound tourists can act as outward manifestations of a country's values and attitude towards the rest of the world, as expressed in their demeanour and behaviour.

Obversely, a host country or region's positive imagery can be constructed to influence inbound visitors' attitudes towards and opinions of that country or region. Thus tourism can be manipulated as a tool of soft power to present an outwardly positive face to the world of a destination country or region which might in other respects be less acceptable. While most explicitly expressed in destinations governed by totalitarian regimes (e.g. Kwek *et al.*, 2014), this facet can be implicit in any (geo)political or commercial (place) branding exercise. Tourism's (potential) recruitment to *Brexit* imagery of a 'Global Britain', as expressed in the *Tourism Action Plan* (DCMS, 2016) (Chapter 5) may be viewed within such a perspective.

Out of this bubbling cauldron of imagery, soft power and nation branding, the concept of 'country reputation' has (re-)emerged, being defined as:

> ... perceptions of a country over time based on direct and indirect experiences with the country as compared to its competitors. (Kiambi, 2017: 62)

That phrase 'over time' would appear to be critical in that while reputations can be destroyed easily and quickly, development of a positive reputation requires much more time to evolve and consolidate. This equally applies to a country's or region's role as a tourist destination. A good or bad reputation can impact significantly on international relations, economic and social transactions, transcending political and economic power. As such, 'reputation' is clearly an element of 'soft power' (Kiambi & Shafer, 2018).

This can be exemplified in the way, even before the 2016 referendum vote, a neoliberal 'Brand UK' was emerging through inward investment flows and the role of consumer product retailing. One study suggested that, such has been the UK retailing reputation (still a nation of shop-keepers-cum-corporations?), that while visitors spend around £4bn annually on shopping when they are in the UK, they may spend four times that amount when back home ordering online from UK retailers' websites (Jary, 2017).

4.7.2 *Brexit's* impact on soft power: How 'global'?

Once Article 50, indicating the UK government's desire to leave the European Union, had been invoked in March 2017, a palpable sense of the UK's diminished status in Europe and beyond could be sensed, at best constraining the government's 'Global Britain' rhetoric (Powell, 2017).

In 1962 former US Secretary of State Dean Acheson pointed out that 'Great Britain has lost an empire and has not yet found a role' (Brinkley, 1990; Deliperi, 2015; Martill, 2017). This was just a few months before French President Charles de Gaulle vetoed the UK's first application to seek a new international role by joining the then European Economic Community (EEC) (Chapter 1). The UK's eventual accession in 1973 did appear to indicate a new identity within Europe, reinforced by the 1975 referendum result that confirmed the country's remaining within the Community. But what of a UK back outside the EU?

Echoing Acheson (and Martill's more recent paper), in November 2017 an article by the *New York Times'* diplomatic editor and former London bureau chief, received wide publicity by arguing that no one now knew what Britain [sic] was or stood for any more (Erlanger, 2017). In the shadow of DJ Trump's elevation to the White House, this article also alluded to former German Chancellor Helmut Schmidt's remark on the UK's interaction with its former colony across the Atlantic, that the supposed 'special relationship' was 'so special that only one side knows it exists' (Marr, 2017: 13). Similar self-delusion appeared to accompany the post-referendum declarations of a new 'Global Britain' that would reinvigorate its relationship with the Commonwealth (Chapter 12).

And thus it was that the European Commission (EC) announced in November 2017 that UK applications for the title of European Capital of Culture (ECoC) 2023 should be discontinued in light of the UK government's invoking Article 50, and following the UK government's repeated intention to withdraw from both the single market and customs union. The EC said it would no longer be possible for a UK city to participate because only countries that were in the EU, the European Economic Area (EEA)/European Free trade Area (EFTA) or in the process of becoming members were eligible for inclusion. This is stipulated in the EU legal text governing the programme (Decision 445/2014, Article 3) written long before the *Brexit* vote was cast.[4]

Because of the variety and complexity of their roles (Richards & Rotariu, 2015), events have developed into an 'industry' in their own right (Getz, 2007). As potential contributions to regional and national imagery and soft power, mega-events such as ECoC hosting can stimulate development, reappraisal and change (Herrero *et al.*, 2006; Richards & Wilson, 2004).

With the UK having been due to have one of its cities designated as European Capital of Culture, along with one in Hungary, for 2023, formal bids, costing hundreds of thousands of pounds, had been submitted in October 2017 by Dundee, Nottingham, Leeds, Milton Keynes and a joint proposal from Belfast and Derry/Londonderry (including Strabane) (Box 4.1).

Despite the stated criteria, the Commission had not previously raised objections to the UK bids. Indeed, under EU law, the UK government had been obliged to launch the competition to find the candidate cities, even after the *Brexit* process was under way, or otherwise face a possible fine.

The programme is eligible to third countries providing they have a clear relationship with the EU, one Scotland and Northern Ireland had wanted to maintain, and all the more disappointing to Dundee (as well as Belfast-Derry/Londonderry) as a claimed frontrunner for the 2023 designation (Robertshaw, 2017).

Only two other UK cities – Glasgow and Liverpool – had previously been recipients of the ECoC title, in 1990 and 2008 respectively. Liverpool estimated that designation had generated a return of more than £750m for the local economy from £170m of spending, fuelling regeneration, tourism and community pride (Carter, 2010). Benefits reported from other past European capitals of culture have included

Box 4.1 Erstwhile UK contenders for European Capital of Culture 2023

Belfast and Derry/Londonderry (http://weare2023.eu)
'People are not contained by boundaries drawn on a map'. The team behind the bid had engaged with 301 artists and 16,000 members of the public.

Dundee (http://dundee2023.eu)
Home to the new V&A Museum of Design. 'We want a young Lithuanian to point to us on a map and say "that's where my friend lives"' as an indicator and celebration of the city's 'truly European' credentials.

Leeds (http://leeds2023.co.uk)
Employing the theme 'Weaving us together' the bid's organisers were aspiring to attract 70% of the city's inhabitants to at least one event, and to increase arts attendance in its five most disadvantaged areas.

Milton Keynes (http://miltonkeynes2023.co.uk)
Employing the tag line 'Different by Design', this city's bid emphasised the quotidian and contemporary, challenging people to 'look again at a place they may think they know, but don't'.

Nottingham (https://www.nottingham2023.co.uk)
The bid team's 'Breaking the Frame' title emphasised the city's rich cultural history, citing such figures as Robin Hood, writers DH Lawrence and Alan Sillitoe, and fashion designer Paul Smith.

Source: Perraudin & Solomon, 2017.

image and identity enhancement, raising local pride and improving local social cohesion, generating international tourist visits and spending, stimulating long-term cultural development and upgrading infrastructure (Dumbrăveanu *et al.*, 2017).

After the Commission ruled that the UK would not be allowed to bid for the designation, the UK government claimed it would be 'working closely with the five UK cities that have submitted bids to help them realise their cultural ambitions ...' (Mason & Walker, 2017). How such ambitions would be realised with reduced local authority budgets and the loss of EU financing was not a question immediately addressed.

As a deflection away from this debacle, much publicity was given to the competition for, and choice of, the UK's own city of culture for 2021. The bid from Coventry was ultimately deemed superior to those from Paisley, Stoke-on-Trent, Sunderland and Swansea. The West Midlands city would be the UK's third city of culture after Derry/Londonderry and Hull (Kingston-upon-Hull),[5] a role modelled on the ECoC concept.

For those cities that do not win such competitions, rather like attempts to establish nation branding, participation has encouraged them to spend huge amounts of money they may ill afford on perhaps futile propaganda programmes. Such actions would appear to arise from the fact that in the 'flurries of apprehension about identity' (Subramanian, 2017: 28), not least expressed in the EU referendum campaign and vote, there may reside the political fear that a city, region or country will be displaced from its erstwhile (self-) recognised economic and/or political role. In relation to the UK's loss of representation within 'Europe', this would appear to be not just a fear, but the reality of diminished soft power, reputation and relevance.

4.8 *Brexit* as (Negative) Mega-event?

Mega-events are distinctive in terms of their large scale, vast impact, international scope and worldwide significance (Hereźniak & Florek, 2018: 93). They have come to be viewed as important in establishing or enhancing identity and place branding. Formerly designed to celebrate particular occasions, mega-events have become increasingly a means to an end rather than an end in themselves: mechanisms for achieving particular goals often related to the branding process. Although not intentionally of 'limited duration', can *Brexit* be conceptualised as a mega-event aimed at reinvigorating/reinterpreting a 'Global Britain' brand, which, in so doing, articulates a spectrum of inverse/negative outcomes?

In any attempt to conceptualise *Brexit* and its likely outcomes, a longer-term view is of course necessary, much longer than this volume is in a position to offer. But sustaining the mega-event model/inverse hypothesis, Table 4.2 attempts to equate Haxton's (1999) six stages of the mega-event experience with the UK's EU withdrawal process.

The hosting of mega-events is intended to generate a wide range of influential benefits both through the actual event itself and its subsequent 'legacy' added-value, benefits which supporters of the UK's

Table 4.2 Parallel stages between mega-events and *Brexit*

	Mega-event process stage	Complementary EU withdrawal stage
1.	Inception	Cameron government decision to hold a referendum
2.	Pre-bid	Referendum campaign
3.	Bid	Referendum
4.	Pre-event	Exit negotiations with EU
5.	Event	Exit
6.	Post-event	Transition period and subsequently

Sources: Haxton, 1999; author's compilation.

Table 4.3 'The mega-event-*Brexit* effect'

Mega-event outcomes	Apparent (pre-) *Brexit* impacts
Financial: revenues for stakeholder organisations	Additional costs have included: small and large businesses having to prepare for an unknown outcome, recruitment of 16,000 extra civil servants, reduction in value of sterling. Financial savings might include a return of contributions made to the EU budget
Economic: increase in spending by visitors and locals	More staycations and a (temporary?) increase in inbound visitor numbers resulting from sterling depreciation
Tourism and international marketing: image promotion and its consequences	Unwelcoming image generated through immigration policy, xenophobic sentiments and a divided society, in contrast to the promoted 'GREAT British Welcome'
Infrastructure enhancement	Reinforcement at borders/ports; expenditure on infrastructure needed elsewhere in the UK deferred because of *Brexit* costs
Urban and regional regeneration	Funding diverted or reduced
Environment: positive impacts	Ambivalent: see Chapter 8
Technological development: supporting sustained economic growth	Economic growth constrained with additional tariffs despite technologically assisted border checks
Human capital: inward skills and knowledge transfers	Outward transfers with the loss of EU workers and FDI
Innovation in collaboration through partnerships	'Global Britain' in bilateral free trade agreements with ...? Loss of EU partnerships
Political capital: advantage for local politicians	What *Brexit* is about for some?
Social and community structure benefits	Loss of funding
Cultural and psychological changes: increased local and regional pride	Divided Britain
Intangibles: such as an enhancement of soft power	Likely loss of soft power

Sources: Matos, 2006; author's compilation.

withdrawal also claimed for *Brexit*. Matos (2006) suggested that there were potentially thirteen elements of such outcomes. Table 4.3 lists these and offers *Brexit* corollaries.

With the exception of the (short-term) increase in numbers of international tourists visiting the UK following sterling's devaluation after the 2016 referendum vote, and the possibility of the return of some monies previously contributed to the EU budget, the (partly speculative) entries in Table 4.3 would suggest that *Brexit* actually or potentially exhibits a negative 'inversion' of the impacts suggested by Matos (2006) for mega-events.

If the referendum's 52-48 vote was symbolic of a divided society, how relevant to tourism and *Brexit* is the literature relating to the role of citizen participation in the generation of imagery and branding (for a 'Global Britain')? The image of a nation, particularly if national attitudes appear fundamentally divided, may defy the clarity implied in such a term as brand image (Ren & Blichfeldt, 2010). Jago *et al.* (2003), for example, argued that events having a strong support from the community are more effective in image creation. But in situations where stakeholders may hold equally contrasting views, the ability to generate a coherent image and to construct a brand that can attract empathy and support would appear to be severely constrained. How far might the image of a divided society, however supposedly 'Global', be able to exhibit a welcoming attraction for potential visitors?

4.9 Conclusion

This chapter has offered a number of potential approaches to theorising the *Brexit*-tourism nexus to help inform subsequent chapters of the book. It would seem that no one theoretical frame is adequate to encompass the multi-layered processes surrounding the UK's exit from the EU and the tourism-related consequences of those. Notably, 'Taking back control' and a 'Global Britain' have 'branded' a post-*Brexit* UK with potentially conflicting images that may enhance or constrain any soft power the country may be able to retain, and the implications for tourism arising from that.

Notes

(1)   And at the time of writing the province had survived without a devolved administration since January 2017.
(2)   The outspoken one-time MP for Southend in Essex: the faded coastal resort with the world's longest pier.
(3)   Qualities recognised by Kotler *et al.* (1993: 18) as 'livability, investibility and visitability'.
(4)   http://eur-lex.europa.eu/legal-content/en/txt.html/uri?=celex:32014d0445&from=en.
(5)   For such a claimed 'outward looking international city', Hull returned a referendum vote in June 2016 of 68% in favour of leaving the EU.

Section B
Tourism Impacts

5 Impact Assessments and Perceptions

Before addressing tourism supply and demand issues arising from UK EU withdrawal in Chapters 6 and 7, this chapter examines the UK government and sector assessments of likely tourism-related impacts that were published in the months prior to the UK's exit. It highlights the limited preparation that had been undertaken to manage such potential impacts.

5.1 UK Government Impact Perspectives

Following the government's change of leadership in the aftermath of the referendum vote, House of Commons Briefing Paper 07213, *EU Exit: Impact in Key UK Policy Areas* (9 June 2015), was reworked as *Brexit: Impact Across Policy Areas* (Miller, 2016). Embracing 19 sectors in 197 pages, the revised document failed to address tourism by name, but embraced several policy areas relevant to tourism, ranging from immigration and the labour market (recognising that accommodation and food services had the highest sectoral proportion of non-UK EU labour at 13%), through transport, citizens' rights, border controls and student mobility to sport, consumer policy and foreign relations. As a background paper, it provided context but little in the way of firm assessment (Table 5.1).

5.1.1 Obfuscation

In a television interview in June 2017, David Davis, then Secretary of State for Exiting the EU and figurehead negotiator with Brussels, said that 'nearly 60' sector analyses had been completed by the government in preparation for EU withdrawal. In the following October Davis told the House of Commons Select Committee for Exiting the EU (*Brexit* committee) that the prime minister had read summaries of impact assessments which, he said, went into 'excruciating detail' (Walker & Mason, 2017).

But when pressed to disclose these, Davis argued that releasing such studies on the likely consequences of different departure scenarios would include: (a) details likely to hamper or undermine the UK's negotiating position with the EU; and (b) market-sensitive information about specific

Table 5.1 *Brexit*: the UK government's background paper to likely sectoral impacts: tourism-related policy areas

Policy area	Detail
Trade	In 2015 the EU accounted for 44% of UK goods and services exports (£222bn) and 53% of UK imports (£291bn): a deficit on trade in goods and services of £69bn, while the surplus with non-EU countries was £30bn. Share of UK exports to the EU had been declining. The UK's new trading relationship with the EU would be the product of negotiation, with a wide range of possibilities.
Foreign direct investment (FDI)	Access to the single market was an important determinant of FDI but by no means the only one. Outside the EU, the UK might be able to establish a regulatory regime more favourable to overseas investors, which could offset the effect of its departure.
Immigration	Possible changes to immigration rules were likely to affect businesses and the economy if it became more difficult to recruit workers from other EU/EEA countries. One possibility would be to restrict economic migration to high-skilled migrants (via a points-based system), thereby reducing the flow of migrant workers undertaking low-skilled jobs (such as in tourism and hospitality). A key question was the extent to which the UK might remain bound by EU free movement laws. The legal status of UK and EU expats would be a significant issue. Leaving the EU did not automatically affect the UK's border controls in northern France, which are based on a bilateral treaty between the two countries.
EU funding	Although the UK had been a net contributor to the EU, certain regions where living standards were relatively low received significant levels of support from the EU budget through the European Regional Development Fund (ERDF) and the European Social Fund (ESF), boosted by matched funding from government or the private sector. Withdrawal would leave a policy vacuum which the government would have to fill to avoid certain regions and sectors losing out.
Employment	A substantial component of UK employment law was based on EU law. In some cases new categories of employment rights had been transposed into domestic law to comply with emerging EU obligations. Some of these had been resisted by the UK government during EU negotiations (e.g. agency workers' rights and limitations on working time).
Agriculture	Brexit, in all scenarios, would mean a departure from the Common Agricultural Policy (CAP), its subsidy and regulatory regime. EU farm subsidies had contributed 50–60% of UK farm income. The UK government would guarantee existing levels of direct subsidies to 2020 'as part of the transition to new domestic arrangements'. It was not clear what levels of support the UK Government would be willing to provide beyond this, or whether it would target subsidies in a different way, for example requiring more 'public goods' in return for any support, such as environmental protection, which the UK government had viewed as the overarching market failure in this sector. Environmental groups were concerned about the overall level of funding for agri-environment schemes outside CAP and how far future UK agriculture policy would support environmental goals. Future access to migrant labour, and how imports would be regulated were critical issues.
Environment	This was an area in which UK and EU law had become highly entwined. There could be fewer incentives for the UK government to meet environmental standards if EU enforcement mechanisms no longer applied. The government would still have certain international environmental commitments and some EU standards. Future governments might decide to increase standards in some areas. A particular challenge could be ensuring effective ongoing coordination with other countries, as many environmental challenges were not able to be tackled in isolation. New mechanisms for coordinating with the EU and between the four nations of the UK could be needed.

Energy and climate change	The Department for Energy and Climate Change (DECC) had confirmed commitment to ratifying the Paris Agreement. Significantly, however, in July 2016 the new government announced that the DECC would be incorporated into a new Department for Business, Energy and Industrial Strategy (DBEIS) [thereby explicitly downgrading climate change prioritisation].
Transport	Much would depend upon whether the UK remained a part of the EEA or whether it concluded bilateral treaties which obliged it to apply much of the current framework regarding single transport markets (as Switzerland had). Airlines wanted the government to negotiate continuing access to the single aviation market. The most obvious way of doing this would be to become part of the European Common Aviation Area (ECAA). The UK's inability or unwillingness to replicate existing market access arrangements could potentially lead to higher air fares. Transport prices could be affected by the general economic impact of Brexit – for example if inflation rose so would rail fares; and if the economy experienced a downturn, large infrastructure projects could prove more difficult to finance.
Human rights	Withdrawing from the EU did not mean withdrawing from the separate European Convention on Human Rights (ECHR), although the government was planning a UK Bill of Rights.
Health	If the UK remained in the EEA it could continue to participate in the EHIC scheme, or, subject to negotiation with EU member states, participate on a similar basis to Switzerland.
Higher education	The loss of funding for EU students could have an impact on the numbers of such students coming to study in the UK, which could have a detrimental impact on fee income, culture and diversity of universities. The UK could lose access to EU research funding, and there were concerns that the movement of high calibre staff and researchers could be impacted, detrimentally affecting the quality of research projects.
Consumer policy	A huge amount of UK consumer protection regulation was derived from the EU.
Foreign and defence policy	Acting through the EU had meant a larger aid budget, the promise of access to the largest consumer market in the world, and a louder political voice. If the UK no longer co-ordinated its policy with member states, it would lose access to these shared 'soft power' tools. However, many UK actions have been taken in conjunction with the US rather than the EU.
The devolved legislatures	With Brexit there could be further policy and legislative divergence in areas of devolved competence, as the UK government and devolved administrations would no longer be required to implement the common requirements of EU directives. This would probably be particularly noticeable in policy areas such as the environment, agriculture and fisheries.
Scotland	Had benefited from both pre-allocated and competitive European funds over the previous four decades. Between 2014 and 2020 Scotland would benefit from a further €4.6bn.
Wales	Had enjoyed access to considerable funding opportunities from the EU, notably from the CAP and Structural Funds. For 2014 – 2020 Wales would receive £1.8bn European Structural Funds investment: with match funding, this would amount to at least £2.7bn.
Northern Ireland	Had benefitted significantly from EU funding: €1.2bn in EU Regional Policy Funding for 2014–2020. UK withdrawal would result in an EU external border running through the island of Ireland.

Source: Miller, 2016.

companies (such as possible undisclosed support for foreign manu-
facturers to stay in the UK). Davis claimed this stance was being taken
because he had 'received no assurances from the [House of Commons
Brexit] committee regarding how any information passed will be used'
(Asthana, 2017: 7). Davis risked being in contempt of Parliament and of
misleading the House of Commons.

The European Parliament (2017–18) had begun publishing a suite of
more than 50 impact assessment studies some months earlier.

Mid-October 2017 saw the House of Commons voting unanimously
(Conservatives taking no part) to demand that Davis's department
provide for the *Brexit* committee the '58' impact assessments – 'secret
advice' – that it had been gathering. Failure to disclose such advice was
preventing MPs from holding ministers to account, it was claimed.
The Department for Exiting the EU had refused even to confirm which
sectors had been examined in the assessments, only responding that
impact studies would be published in 'due course'. This suggested
that the government was withholding inconvenient truths about likely
economic and social impacts of the UK's EU withdrawal.

In early November 2017 the UK government conceded that it must
release the previously confidential documents to the House of Commons
Brexit committee for deciding which elements could be released more
widely.

The minister then claimed that such information did not exist in the
form of impact assessments. Eventually, the committee was handed two
files containing 850 pages of mainly background sectoral analyses. This
was insufficient for some committee members. Called to the committee,
Davis argued that impact assessments would be completed 'a little closer
to the negotiating timetable'. Contingency planning studies had covered
such areas as customs and aviation, but 'don't have numbers attached
to them' and so, according to Davis, did not constitute a forecast.
He had been misunderstood, he claimed. The minister also claimed
that government plans for the possibility of leaving the EU without
agreement did not include any impact assessment (Walker & Mason,
2017: 14–15).

It had been reported previously that civil servants in Davis's
department had been so alarmed about the lack of preparation for a
no-deal exit that they had taken to writing emails stressing the dangers,
in order to cover their backs for any future inquiry (Hyde, 2017).[1]

5.2 'Tourism Action Plan'

The British people's decision to leave the European Union creates many
great opportunities for growth, such as cutting red tape and forging partnerships in new and developing markets. Our stunning scenery hasn't
changed, nor our centuries-old monuments and cultural traditions.

Together with industry, the Government will work to ensure that tourism continues to thrive as negotiations on the UK's exit progress. The end goal is a Britain that is even more attractive, accessible and welcoming to visitors. (May in DCMS, 2016: 2)

As a barely veiled apologia for the likely impacts of the UK's withdrawing from the European Union, the government's *Tourism Action Plan* (DCMS, 2016), was published just two months after the EU referendum, and almost simultaneously with the Impacts briefing paper discussed above (section 5.1). It was short on new thinking but full of hyperbole and rhetoric.

Tourism strategies for Scotland and Wales were also 'refreshed' during the year (VisitScotland, 2016; Welsh Government, 2016).

As ever, the Westminster document claimed that tourism was one of the UK's most important 'industries' directly responsible for 1.6m jobs with tourism-related sectors employing a total of 3.1m, representing 'approximately' 1 in 10 jobs in the UK.[2] It recognised, however, that over 50% of international visitor spend took place in the capital, such that 'there is more to be done to rebalance the sector, boosting jobs and growth right across the country' (DCMS, 2016: 3). Was this also a conscious response to the regional geography of the referendum vote?

The *Action Plan* aimed to show how the government was 'working to do this [rebalancing and boosting] through action' in relation to five specified areas:

  (i)   The tourism landscape: strengthening co-ordination and collaboration
  (ii) Skills: boosting apprenticeships and attracting more people to careers in tourism
  (iii) Common sense regulation: examining the scope for deregulation
  (iv) Transport: making it easier for visitors to explore by rail, bus and coach
  (v)   A GREAT Welcome: driving continuous improvements in our visa service ...

The actions in this report together with our new industrial strategy will make our tourism sector more internationally competitive and resilient and ensure that its growth delivers for everyone. (DCMS, 2016: 3)

These five areas are addressed below.

5.2.1 'Tourism landscape'

Given its role as a post-*Brexit* tourism promotion framework for the UK, the actions in this section of the *Tourism Action Plan* needed to be viewed through the perspective of (i) reduced funding for local authorities and destination marketing organisations (DMOs), from both the

UK government and the EU, (ii) the centralisation of *VisitBritain*, (iii) the potential of Scottish independence and (iv) potential discontinuity in Northern Ireland.

Noting that the costs of collaboration could be high because of the sector's diversity, fragmentation and competitiveness, the government was 'taking steps to address this' constraint by 'creating an overarching industrial strategy', apparently discovering that '[i]t is important to ensure that policy is considered in the round and action is joined up' (DCMS, 2016: 6).

The *Action Plan* articulated the multiple sources – fragmentation? – of policy influencing tourism development and promotion by pointing to the several government departments that were stakeholders in the success of Britain's tourism sector. Strengthening marketing and promotion to overseas visitors, and particularly those from long-haul destinations, was clearly a requirement born of the fear that the high proportion of international tourists visiting the UK from other EU member states could not be assumed after whatever form of EU exit the UK took.

As specific exemplars of coordinated promotion, discussions of the Tourism Industry Council would inform the UK government's negotiating position on EU withdrawal (and in relation to development of the wider industrial strategy).

Industry support partnerships would be enhanced (VisitBritain, 2015, 2016b), including a three-year £10.2m joint venture between *Expedia* and *VisitBritain* to market the UK to the USA and EU member states Germany and France, the UK's three largest and most valuable inbound visitor markets. Digital marketing through bespoke content platforms was a high priority in this campaign (VisitBritain, 2016b).

An event support programme[3] had been put into place, overseen by an Events Industry Board[4] (advising government on the implementation of the UK's business visits and events strategy: DCMS, 2013a), to scrutinise applications for financial and advocacy support for UK cities and companies bidding to host international events in the UK: a sector worth over £41bn in direct visitor spend (Tourism Alliance, 2017: 14).

Aimed at raising the value and volume of international events held in the UK, within a year Tourism Alliance (2017: 14) was claiming that the benefits from a coordinated cross-government approach to bidding and staging events and conferences in the UK were becoming apparent.

'Events produce a diffuse and beneficial effect far beyond increased visitor spend by delegates in the immediate geographical area' (DCMS, 2016: 7). That is, by projecting the kind of soft power that was being diminished with the UK's exit from the EU, as evidenced in the European Capital of Culture debacle (section 4.7 above).

'Professional, creative and research networks are strengthened by gathering for events, and an event can also create a powerful association between the industry and the host location, giving local businesses and organisations in that sector a competitive advantage' (DCMS, 2016: 7). This would be in the wake of the research networks and professional relationships being lost by the UK's exit from EU programmes.

To help address the imbalance of tourists visiting only London, a £40m Discover England fund was launched to encourage overseas tourists to travel beyond the capital. The plan sought to encourage new tourism developments outside London, ease travel around *England*, and help inbound visitors as well as domestic travellers to explore tourism opportunities. Initial grants would be awarded to projects and organisations in different regions. Not a little ironically, winning bids would 'ideally align investment in tourism and culture from a range of different private and public funding streams' including the EU LEADER programme, which, in 2016, provided funding to local businesses through 80 LEADER action groups across the UK. The £40m government support would be spread over three years: £6m April 2016– March 2017; £12m 2017–18; £22m 2018–19, the latter neatly coinciding with the UK's putative final year as an EU member state. Project evaluation would continue into 2020 (Balcombe, 2016).

Domestic tourism comprises over 60% of total tourism revenue and, aside from government expenditure, represents the most significant mechanism for redistributing economic activity from urban to rural and coastal areas. If government was to achieve the goal of further spreading the benefits of tourism to the regions, destination management organisations (DMOs) needed a viable mechanism for accessing the long-term sustainable funding required to develop and promote the high quality world-class tourism products trumpeted in the *Tourism Action Plan*.

Yet, since 2008, public funding for domestic tourism support had been reduced from £197m to just £71m, resulting from the abolition of regional development agencies (RDAs), reductions in local authority funding, and the 'rationalisation' of *VisitEngland* and *VisitBritain*. Hopes that the private sector would help to compensate for the funding reduction had proved misplaced. With DMOs having neither a statutory remit nor access to stable funding, fragmentation and income chasing had resulted (Tourism Alliance, 2017: 6).

5.2.2 'Jobs and Skills'

In responding to a combination of specific local and regional needs and the potential loss of large numbers of EU workers in tourism and hospitality, food processing and agricultural production, the 'Jobs and Skills' section of the *Tourism Action Plan* aimed at boosting apprenticeships and attracting more (indigenous?) young people to careers in UK tourism and hospitality. With the objective to attain 3 million apprenticeship starts in *England* by 2020, a pilot apprenticeship scheme recognising 'that the seasonal nature of tourism can act as a barrier to smaller and rural businesses' (DCMS, 2016: 9) was to enable apprentices to complete their training over 16–18 months instead of 12, with a mid-break to permit their undertaking other activities. Partner companies in this scheme included Monarch Airlines which went out of business the following year.

No reference was made in the document concerning the monitoring and appraisal of the quality and achieved objectives of these schemes.

To fund the three million new apprenticeships, a levy was introduced to raise £2.5bn a year by 2019–20, payable by any company with a wage bill over £3m. Most firms were not large enough to be liable for the levy. However, medium-sized enterprises with wage bills under £3m that employed between 50 and 200 staff, were faced with new responsibilities including releasing apprentices for one day a week for off-site training, and contributing 10% of the training costs. This made apprenticeships less popular with employers. As a result, there was a 59% drop in the number of apprenticeship starts by the end of the first academic year following the scheme's introduction (Bloom & Hoggan, 2017).

With claimed employment in tourism growing at a faster rate than total UK employment, a stated aim of the *Tourism Action Plan* was to improve perceptions of the tourism sector particularly for young people. The #*mytourismjob* campaign sought to encourage more young people to consider tourism and hospitality as a career option. This would be alongside efforts of the Careers and Enterprise Company, established in 2015 to help link schools and colleges to employers (Careers and Enterprise Company, 2018; see also Offord, 2018), to assist schools in developing a tourism dimension in their career planning (DCMS, 2016: 9). Certainly, there needed to be greater provision of careers information, advice and guidance that reflected true career opportunities in the sector (Tourism Alliance, 2017: 9).

According to the Institute for Fiscal Studies (IFS) (Amin-Smith *et al.*, 2017), while there was a 'desperate need' for better vocational training, the government's target to rapidly increase the number of apprentices risked being poor value for money, showed a 'cavalier use of statistics' and could devalue the 'brand' of apprenticeships by turning it into 'just another term for training' (Coughlan, 2017).

If the government's 2017 industrial strategy was to complement the *Tourism Action Plan*, it was important that tourism and hospitality be included in any priority funding. Such funding needed to incentivise the delivery of sector programmes in full-time provision and sector apprenticeships (Tourism Alliance, 2017).

The industrial strategy was built on 10 'pillars' that needed to be addressed in order to assist increased growth (UK Government, 2017a). The pillar on 'developing skills' highlighted the weakness of technical education, and promoted the government's post-16 skills plan as a solution. But with reduced funding to further education and a weak careers guidance structure it was unlikely to meet the government's objectives and help tourism and hospitality sector businesses.

In the pillar on 'creating the right institutions to bring together sectors and places' the *Tourism Action Plan* was highlighted as an example of government working closely with a particular sector. Yet the *Action Plan*'s plan strand on skills and jobs did not address such fundamental issues as recruitment difficulties and changing employee expectations.

Much of the industrial strategy's focus was on STEM (science, technology, engineering and mathematics) sectors, because this was where it believed skill shortages, and the UK's competitive position, lay. The tourism and hospitality sector could only assume that this would be to its detriment (Kent, 2017).

Muted support for rural tourism, or rather, for the ability of small businesses to operate from rural areas, was expressed in the *Tourism Action Plan*. Comprehensive access to effective, efficient and affordable broadband was essential to the successful operation of rural businesses, and the creation of a universal service obligation of at least 10Mbps that imposed a legal obligation on the provider was necessary. This requirement was unlikely to be diminished by *Brexit* (Tourism Alliance, 2017: 7).

5.2.3 'Commonsense regulation'

The *Tourism Action Plan* claimed that, from a business perspective, successive governments had failed to reduce sufficiently the level of regulatory 'burden' faced by the tourism sector, despite such initiatives as the Red Tape Challenge (Cabinet Office, 2011). Few regulations had been removed compared to the number of new ones introduced. The government had been carrying out a review of local authority regulation and enforcement 'to find ways of cutting red tape and reducing bureaucratic barriers to growth and productivity' (DCMS, 2016: 11), although this could represent a further hollowing out of local government functions.

The EU's revision of Package Travel Regulations (PTRs) from April 2018 provided such benefits for travellers between the UK and the EU as ensuring repatriation if their tour operator failed, and providing legal recourse for customers in their home country. While the revised PTRs would remain important after *Brexit*, such regulations had long been problematic for the domestic tourism industry as they had been seen to constrain small, local businesses from working together to develop value-added products (those that combined just accommodation and a service other than transport) (Tourism Alliance, 2017: 10–11).

Tourism Alliance (2017) argued that leaving the EU allowed the UK government to resolve this problem by making the provision of transport a requirement of any domestic travel package. This would apply to businesses conducting tours but not to those where the visitor was undertaking the journey themselves and simply buying a value-added product such as a weekend at a B&B that included a meal at a local pub or tickets to a local attraction.

In recent years the arrival and rapid growth of sharing economy businesses that operated on the margins of regulatory compliance had exposed holes in regulatory controls (which the government and Tourism Concern appeared to wish to reduce further). Sharing economy facilitators had been able to gain significant competitive advantages over existing businesses, a problem exacerbated by local authorities' severely reduced funding that inhibited their ability to enforce compliance. A review of the appropriateness of accommodation-related regulation was required.

5.2.4 'Transport'

> Britain's road system is an excellent means of exploring the country. However, many international visitors are fearful of driving on the 'wrong' side of the road, so it's vital that they feel confident navigating our system of public transport, if they are to explore beyond London. (DCMS, 2016: 12)

This rather lame opening to the *Tourism Action Plan*'s section on tourism transport signalled again the urgent need to diffuse visitor numbers and spending in a more balanced way across the UK. It belied the often inchoate and unconnected nature of much of Britain's [sic] deregulated and privatised rail and road public transport provision.[5] That '… international visitors can find it difficult to find the information they need, in the right language, to book the right tickets at the right price, and to be assured of a seamless journey' (DCMS, 2016: 12) was an experience by no means confined to international visitors.

It reflected the failings of successive governments: in extolling the 'virtues' of competition rather than recognising travellers' requirement for intra- and multi-mode coordination; twice re-privatising the east coast main railway line after it proved eminently profitable for the taxpayer in state hands; and shelving electrification plans for lines in the south-west and north of England, and modernisation in Wales, Yorkshire, Cumbria and the English Midlands, in contrast to the enormous sums (£17bn+) being spent on the Crossrail project in London.

Following the referendum vote, the Chancellor had promised an extra £300m for train services in northern England, which was later reneged upon. By contrast again, the High Speed 2 rail line, largely benefitting southern Britain and exacerbating regional imbalance, would cost £400m *per mile* (Harris, 2017b; Pendleton *et al.*, 2019).

Tourism Alliance (2017: 12-13) pointed out that travel was the essence of tourism, with 1.65bn domestic trips and 101m international trips being undertaken in the UK each year. There needed to be a stronger public transport sector – more efficient, better value and better integrated – and significant investment in UK transport infrastructure (especially away from London). *Brexit* provided a number of opportunities for enhancing the role of both air and surface transport in UK tourism (Box 5.1). The extent to which these were consonant with the government's required environmental goals was questionable.

Only just over 20% of people used public transport to undertake tourism-related travel in the UK: much more needed to be done to promote and support the use of rail and public road services and cycling.

Box 5.1 Tourism Alliance suggestions to enhance the role of air transport in UK tourism after *Brexit*

- Alleviate congestion in south east England, and render regional airports across the UK more attractive for both international and domestic travellers. The latter could be assisted through the UK's regional air connectivity funds, which, arguably, had been hampered by the EU's state aid rules restricting such funds' use to route development at airports that had fewer than 5m passengers a year.
- Abolish air passenger duty (APD) on domestic travel, and preferably within an overall policy of removing APD from all routes – domestic, European and international.
- Modernise the UK's airspace: such modernisation was urgently required to ensure capacity kept pace with demand. Without it, delays faced by passengers could rise to 4m minutes by 2030.
- Improve surface access connectivity between airports and tourists' final destinations, both through improved infrastructure and services, and better integrated transport modes.
- Link together the regional connectivity funds with such strategies as the Discover England Fund and the GREAT campaign. This could be achieved through an integrated programme whereby products developed through the Discover England Fund were sold in markets accessed by routes started by the Regional Connectivity Fund and were promoted in those markets through the GREAT campaign.

Source: Tourism Alliance, 2017: 12-13; see also Butcher, 2016; Gardner and Kries, 2017

More proactive encouragement including further financial incentives for the development of electric vehicles, their infrastructure and use, was crucial.

5.2.5 'GREAT Welcome'

VisitBritain/VisitEngland's 'GREAT Welcome' programme, as part of its 'GREAT Britain' marketing theme, had pursued the 'Global Britain' ethos of looking well beyond Europe for new and expanding markets, including improving the visa system for 'key' tourism markets that included the Arabian Gulf and China (DCMS, 2016: 14) (see Chapter 13).

The GREAT Campaign appeared to be well supported by the UK tourism sector, but was seen to need enhancement particularly to address issues of image perception and to boost business tourism from new markets. Funding for *VisitBritain* needed to be raised to provide increased opportunities for both strategic and tactical marketing of the UK as a tourism destination within its 'new' 'global' circumstances (Tourism Alliance, 2017: 14–15).

Since biometric visas were introduced in 2008, the UK's share of outbound tourism from China, Russia and India had actually decreased by 30%. The Home Office needed to work with the tourism sector to develop a coordinated and coherent programme of measures for post-*Brexit* UK to remain a destination of choice for travellers from the BRIC growth markets (Tourism Alliance, 2017: 14–15).

In the meantime, how many European visitors might be lost?

5.2.6 Useful contribution?

In its concluding 'Next steps' section, the *Tourism Action Plan* highlighted that through the GREAT campaign 'We will be show-casing British food and drink, through DEFRA's great British Food Campaign' (DCMS, 2016: 15), despite the imminent loss of EU workers in these sectors (see Chapter 6). This was perhaps an unfortunate note to end on.

The emergence of more corporatised public planning approaches, barely distinguishable from those of the private sector – as clearly expressed in the aspirations of the *Tourism Action Plan* – could be seen to be a result of the emergence of 'new' public management within the neoliberal agenda. They repeatedly emphasised public-private partnerships in tourism planning and development (Hall, 2015; Saarinen *et al.*, 2017), which the UK tourism sector appeared to accept uncritically.

Thus Tourism Alliance's (2017: 4–5) response presumed that the *Action Plan* would be sustained as the UK's key tourism policy document through EU withdrawal and beyond. Its effectiveness to 'grow and spread' the benefits of tourism, however, remained to be seen.

5.3 The UK Tourism and Travel Sector's Perceptions and Responses

Acknowledging the potential for sector upheaval, key visitor economy organisations contracted, or themselves undertook amongst their members, surveys of how the UK's withdrawal from the EU would likely impact on inbound, outbound and domestic tourism processes in the short- and longer-term. Deloitte (2016) for ABTA, and Tourism Alliance (2016, 2017; Janson, 2016a, 2016b) conducted such surveys before and after the EU referendum to identify the most significant areas of anticipated impact. Mass- and social media also offered sometimes less informed opinions and perceptions.

As an academic response, in his early post-referendum content analysis of 20 mostly newspaper articles from around the world, Lim (2018) identified 10 emergent themes for likely *Brexit* impacts on tourism: bilateral agreement; border control; travel protection; duty-free allowance; foreign exchange rate; travel costs; past travel experience; need to travel; travel intentions; and actual travel.

In a pre-referendum study of travellers' perceptions, global travel publisher *Travelzoo* and Bournemouth University (Travelzoo, 2016) claimed that *Brexit* could cost the UK's tourism industry up to £4.1bn a year in lost international tourist spending. This was based on survey findings that around a third of Italian (33%), Spanish (33%) and German (30%) travellers, and a quarter of those from France (24%) said they would be *less inclined* to travel to the UK in the event of a 'Leave' vote. Four in 10 respondents from EU countries were also concerned that EU withdrawal could make UK holidays more expensive (Davies, 2016). Even respondents from further afield – 10% of those from Canada and 12% from the US – stated they would be less likely to visit a post-*Brexit* UK.

Holidays for Britons in Europe could become more expensive in less obvious ways. Forty percent of French and Spanish respondents felt it would be fair to impose higher fees, such as city taxes, on UK visitors as non-EU members.

Of UK travellers surveyed, 28% were concerned that withdrawal could lead to more expensive holidays, while 56% were concerned that *Brexit* would reduce the ease and flexibility with which UK nationals had been able to travel within the EU.

Other UK concerns included: the price of holiday insurance and the cost of mobile roaming, while 22% were concerned that UK beaches could become more polluted without strict regulations enforced by the EU (Travelzoo, 2016).

Similar findings emerged from a World Travel Market (WTM) survey of a thousand UK travellers, where 54% expressed concern about the state of sterling and increased holiday costs, 33% were fearful of the heightened risk of passport control queues, while 24% claimed to hold no concerns about *Brexit* and its impact (WTM, 2017).

Of the *Travelzoo* sample surveyed who were in favour of an 'independent' UK, 61% said they would be willing to pay more for their holidays to be out of the EU (Davies, 2016; Travelzoo, 2016).

5.3.1 Stakeholders' categories of concern

Most debates within the sector revolved around what EU rules and regulations would need to be renegotiated or replaced and how these were likely to impact on customers and businesses. Deloitte (2016) for ABTA identified four categories of potential impacts (Table 5.2).

Tourism Alliance identified nine main areas of impact, as indicated in Table 5.3. This reflected the outcome of a survey conducted during August and September 2016 of 61 Tourism Alliance members and associated businesses (Janson, 2016a, reported in Sima, 2017).

While almost 64% of Tourism Alliance respondents considered that changes to border controls were likely to have negative or very negative impacts on tourism, approaching a third (31.2%) considered there would be no impact, perhaps assuming the UK would remain within the single market.

From these findings it was claimed that the tourism sector was unanimous in believing that with the UK's exit from the EU customs union:

- any imposed border checks or levies should be kept to an absolute minimum;
- a comprehensive free trade agreement be struck with the EU in the shortest possible timeframe;
- electronic trading arrangements be developed wherever practicable; and

Table 5.2 Deloitte/ABTA summary of potential *Brexit* impacts

Impact areas	Impact categories
Consumers	Roaming fees
	Package Travel Directive
	Freedom of movement – visa restrictions
	European Health Insurance Card
	Consumer Rights Directive
	Passenger rights
Relevant employment	Service quality
	Skills shortage
Investment and currency	Unstable pound
	High risk investment factor
Business and consumer confidence	Negative impacts on business confidence
	Higher production costs leading to higher selling prices
	Negative impacts on consumer spending

Source: Sima, 2017, after Deloitte, 2016.

Table 5.3 Tourism Alliance survey: perceptions of potential impacts (% of sample)

Main areas of impact	Very negative impact	Some negative impact	No impact	Some positive impact	Very positive impact
Introduction of tariffs and custom charges	32.8	32.8	34.4	0.0	0.0
Introduction of non-tariff barriers	27.9	31.2	36.1	3.3	1.6
Divergence from European Union (EU) on regulations and standards	23.0	31.2	19.7	18.0	8.2
Reduced access to EU programmes	41.0	34.4	24.6	0.0	0.0
Tightening of UK immigration policy for EU nationals	27.9	45.9	21.3	1.6	3.3
Withdrawal of the UK from EU travel agreements	37.7	36.1	21.3	3.3	1.6
Changes to border controls	42.6	21.3	31.2	1.6	3.3
Divergence from EU on consumer protection	23.0	32.8	32.8	8.2	3.3
Restrictions on UK nationals working in the EU	26.2	24.6	47.5	1.6	0.0

Sources: Janson, 2016a; reworked from Sima, 2017.

- transitional arrangements were such that UK businesses did not face prohibitive tariffs when exporting or importing goods and services in any period between the UK exiting the EU and the signing of a future trading agreement (Tourism Alliance, 2017: 12–13).

The likelihood of visa requirements for travel between the UK and EU countries – dismissed by some – was a significant area of concern. Having to apply for a UK visa could discourage EU nationals from visiting the UK. Potentially, this could significantly impact upon inbound tourism to the UK, given that the EU has been the main source of overseas tourists coming to the UK and the main overseas destination for UK tourists (ONS, 2019b: Tables 2.10, 3.10).

Almost three-quarters – 73.8% – of Tourism Alliance respondents identified the tightening of UK immigration policy for EU nationals as having likely negative or very negative impacts for inbound and domestic UK tourism (Table 5.3). Even with an open immigration policy the unemployment rate in the UK had been steadily falling, and demand for qualified tourism, hospitality, events and aviation staff had been increasing (Sima, 2017) (see Chapter 6).

KPMG (2017) drew up its own analyses of sixteen broad sectors in terms of their sensitivity to access to EU workers and dependency on export markets, and identified food and drink (including farm workers), and hotels and restaurants as two of the three sectors most exposed to a loss of EU labour. Labour for transport was also an area of concern. Failure to reach a positive EU withdrawal agreement would have a disproportionately severe impact on food retailers (with knock-on cost effects for the hospitality sector and for domestic consumers) because WTO tariffs on imports would be highest for such staples. The highly interconnected nature of UK and EU supply chains was crucially important for many businesses.

Three-quarters – 75.4% – of Tourism Alliance interviewees envisaged likely negative impacts following a loss of access to EU programmes (Janson, 2016a). European Union funding had been employed in the development and promotional efforts of many UK tourism destinations, attractions, museums, organisations and programmes (see section 6.5).

Several intangible impacts arising from *Brexit* were acknowledged in the survey responses, notably uncertainty, lack of trust and negative destination image. Such impacts were reflected in, and often generated by, the UK's mass- and social media. These suggested, at the very least, the UK's loss of soft power.

5.4 Impact on Business Confidence

While businesses were aware that they needed to prepare for the UK's EU withdrawal, for too long there had been too many uncertainties to be able to do so effectively. The appetite for risk declined and more companies adopted defensive strategies. In such an atmosphere of continuing uncertainty, organisations wanted to avoid unnecessary effort or costs of planning for an unknown settlement. Thus, for example, in April 2018, a survey of company board members found 51% of UK firms surveyed had begun to prepare for a 'no deal' *Brexit*. Contingency plans included non-UK subsidiaries switching to EU-based suppliers (35%), reducing investment in the UK (23%), and moving job roles (15%) or operations (14%) out of the UK. A further 40% planned to inaugurate *Brexit* contingency plans by the end of 2018 if no trade deal or transitional arrangements had been agreed (Pinsent Masons, 2018).

One Brussels-based think-tank claimed the value of assets about to be transferred to 'mainland' Europe could total £1.8tn, equivalent to 17% of the UK's banking assets, while an LSE study estimated the potential loss of business revenue to the UK exchequer of such moves at 15% (Harris, 2017a). Business and conference tourism could suffer considerably as a result.

5.5 Conclusion: 'Managing' Impacts

Tourism Alliance surveys concluded that successful *Brexit* impact management needed to focus on:

(1) clarifying the UK's position in relation to EU funding: ideally the UK could both contribute and benefit from shared financial and know-how resources;

(2) consolidating the roles of EU workers while at the same time discouraging mass immigration;

(3) increasing promotional activities overseas to minimise negative destination images generated by the nature of the referendum campaign and its aftermath;

(4) clarifying exactly what trade-related agreements the UK wanted with Europe so that business could plan ahead (Janson, 2016a, reported in Sima, 2017).

The most important EU regulations or directives needing to be prioritised in negotiations were perceived by survey respondents as: freedom of movement, VAT rules, the package travel directive, and the working time directive (Janson, 2016a, reported in Sima, 2017).

Agreement on how these might work after withdrawal had not been forthcoming. The sector remained unsighted.

Positively, a post-EU UK would still function in a globalised world, where a major business partner would be the European Union. Therefore, future debates around the long-term impacts of withdrawal for tourism would need to look beyond sterling fluctuations, the price of holidays or the ease with which travellers were able to cross borders, towards building long-term relationships with EU-based partners and navigating new (transposed) UK legislation and regulations in conjunction with managing persisting EU legislation and regulations (Sima, 2017: 297–302).

Notes

(1) For the official version of this protracted process see Maer and Ryan-White (2018).
(2) https://www.gov.uk/government/statistics/dcms-sectors-economic-estimates-2016
(3) https://www.visitbritain.org/events-support-programme-overview
(4) https://www.gov.uk/government/groups/events-industry-board
(5) In Northern Ireland rail and major bus provision remained in the public sector.

6 Supply-side Issues

The focus of debate on tourism supply-side issues impacted by the UK's exit from the EU largely surrounded the hitherto role of EU migrant labour, education and training, transport, food supply, taxation and regulation. These issues were inadequately addressed in the UK government's *Tourism Action Plan* (DCMS, 2016). Building on the sector perceptions articulated in Chapter 5, this chapter briefly assesses the main elements of these issues.

6.1 Tourism and Hospitality Employment

The EU's freedom of movement principle, enshrined in the Treaty for the European Union, has guaranteed the rights of individuals from one EU member state to live, travel within, and work in another member state.

While the UK did not join the EU's border-free Schengen Area, nor agree to share common visa rules and processes, other EU citizens have been able to enter the UK almost without restriction. This has exerted significant direct and indirect impacts on the tourism and hospitality sector which has relied heavily upon non-UK workers.

The EU social chapter of employment and social rights is extensive. For example, the Working Time Directive stipulates maximum working hours and establishes minimum periods of paid leave for contracted workers. While the costs of complying with social regulations were cited in *Brexit* discussions, it was often claimed that employee rights were well established in the UK, and that the government had gone beyond minimum EU rules in many areas. It was, therefore, unclear whether *Brexit* would radically alter the balance of rights and responsibilities held between employers and employees.

Particularly relevant to the outbound travel industry, the EU has ensured consistent rules for employers and employees when operating or working across the EU, as laid down by the Posted Workers Directive and further facilitated by other measures, such as the Recognition of Professional Qualifications Directive. These rules, and similar legislation, simplified operations and enabled cross-border working (Chapter 3).

6.1.1 The dimensions

Migration has had a positive impact on the recruitment of skilled and unskilled workers into the UK tourism and hospitality sector

(Janta & Ladkin, 2013). Particularly in the case of unskilled workers there exists the paradox that while much of the sector remains part-time, seasonal and insecure, it is often for those very reasons that working in tourism, hospitality and events has become highly attractive for the EU migrant workers and students who fill many of these flexible occupations (Hannam, 2017).

The UK's GDP growth slowed during 2017 to rank as the lowest in the G7, and the OECD expected the growth rate to slip further in 2018, keeping the country at the bottom of the growth league (Inman, 2017). Under these circumstances, the contribution of tourism and travel has grown both relatively and absolutely. The free movement of labour within the EU single market has facilitated easy and cost-effective hiring processes and has allowed the sector to increase capacity quickly in peak demand times. The sector is (variously) claimed to generate 4.2m jobs in total (Deloitte, 2016: 12), with 2.13m people being directly employed in the UK (People 1st·, 2017a).

It was estimated that between 2014 and 2024, the UK tourism and hospitality sector needed to recruit 1.3m staff (75% of whom would be needed simply to compensate for staff turnover: a figure which, itself, raises serious questions). Staff engagement, better retention rates and career progression would pose major challenges for the sector (People 1st·, 2017a).

At the time of the EU referendum there were some 5 million non-British nationals working in the UK, accounting for 17% of the total workforce. Across tourism and hospitality, 24% of the workforce was made up of non-British nationals, compared to 16% in retail and 21% in the care professions (People 1st·, 2017a). In 2016, the proportion of migrant workers in tourism and travel coming from other EU countries comprised 48% in hospitality, 44% in aviation, 39% in tourism and 38% in retail (People 1st·, 2017b): and over the previous five years the percentage increase in EU labour in these sectors had been significant (Table 6.1). Yet it was estimated that post-*Brexit* 96% of EU workers in the UK hospitality sector would not be eligible to gain entry under the UK's non-EU immigration system because of the sector's relatively low skill and pay levels (British Hospitality Association, 2017; CBI, 2018).

As UK unemployment fell from over 8% in 2011 to below 5% in 2016, most sectors saw an increasing dependence on migrant workers, both nationally and across all devolved nations (Table 6.2) and English

Table 6.1 Sub-sectoral growth of EU migrant labour in UK, 2011–2016

Sub-sector	Tourism	Hospitality	Aviation	Retail
% of migrant workers from EU, 2016	39	48	44	38
% increase 2011–2016	48	61	47	46

Source: People 1st., 2017b.

Table 6.2 Dependency on migrant workers in tourism and hospitality occupations across the UK home nations, 2011–2016

Area	Size of the workforce	Projected additional employment 2014–2024	Reported hard-to-fill vacancies %	Reported skill shortages %	Migrant workers %		Of whom EU migrants %	
	2016				2011	2016	2011	2016
England	1,749,830	1,118,590	40	63	24	26	36	47
Scotland	196,884	100,863	43	64	15	20	44	61
Wales	116,456	47,341	47	80	10	11	51	50
Northern Ireland	49,883	27,467	33	60	16	18	–	74

Source: People 1st., 2017a: 4.

Table 6.3 Regional employment and migrant workers in tourism and hospitality occupations across *English* regions, 2016

Area	Size of the workforce	Projected additional employment 2014–2024	% of reported hard-to-fill vacancies	% of reported skill shortages	% of migrant workers	% of EU migrants
North East	80,096	46,748	42	58	10	40
North West	238,595	133,449	37	53	20	46
Yorkshire & Humberside	169,127	76,255	40	67	14	40
East of England	187,388	117,435	37	64	20	49
East Midlands	145,948	67,722	42	60	16	39
West Midlands	154,844	113,395	39	63	19	39
London	291,255	230,729	26	84	63	43
South East	277,589	178,575	42	63	24	63
South West	204,989	154,283	46	49	15	51

Source: People 1st., 2017a: 5.

regions (Table 6.3). Some regions, most notably London, had a particularly high level of dependence on immigrant labour in the sector (People 1st·, 2017a).

Although following a post-referendum fall, net migration appeared to be holding steady, the first quarter of 2018 brought the first significant decline (3% year-on-year) in the rate at which migrants were filling vacancies: a factor particularly noticeable in London, North West England, the West Midlands and Wales (Resolution Foundation, 2018: 2–3). Perhaps in response, hotels and restaurants was recorded as one of the three sectors with highest pay growth. In agriculture, by contrast, pay levels fell in real terms (Elliot & Kollewe, 2018).

By 2017, 38% of hospitality and tourism businesses were reporting hard-to-fill vacancies, a situation being exacerbated as the labour market tightened: 21% of businesses further reported that the staff they currently employed lacked essential skills.[1] This was despite the sector spending £1.2bn on providing training for 1.5m staff in 2016 and supporting over 18,000 sector-specific apprenticeships (Tourism Alliance, 2017: 8–9).

One estimate suggested that the hospitality sector required 62,000 EU migrants per annum in order to be able to maintain existing activities and to grow (KPMG/BHA, 2017).

Amongst the English regions, London, the South West, South East, East of England and the West Midlands were most dependent on non-UK EU nationals to make up their workforce. London was the most dependent on non-British nationals, with 63% of the tourism and hospitality workforce comprised of migrant labour, of whom 43% was from other EU countries in 2016 (Table 6.3). Reported patterns of skill shortages suggested that at this macro level non-British nationals were not necessarily filling skilled vacancies in significant numbers.

The increased reliance on non-UK workers to fill vacancies is high-lighted when looking at the breakdown of sub-sectors. With the exception of restaurants, from 2011 to 2016 all tourism and hospitality sub-sectors saw a substantial percentage increase in the number of EU workers (Table 6.4).

Clearly, any restrictions on UK tourism-related businesses employing EU nationals would place a serious strain upon the sector. Growing staff shortages would result in increased employment costs as businesses competed for fewer (skilled) workers and needed to spend more on training. This would lead to higher prices for consumers, compromising the UK tourism sector's competitiveness (Tourism Alliance, 2017: 8–9). Another contradiction for 'Global Britain'.

The sector would also lose some of the longer term investment it had made in skills if EU employees' working rights were restricted or if they left the UK. Indeed, by the end of 2017 significant numbers of departures of EU staff were being recorded as the result of the fall in the value of sterling, stronger growth in their home economies, and uncertainty over future employment opportunities in the UK (KPMG, 2017: 7).

More than a quarter of inbound visitor arrivals and £5bn per year spending were attributed to the VFR markets of EU nationals working and living in the UK. A decline or reversal of migration flows would lead to a loss of income for accommodation, travel, attractions, food and drink sectors, as well as in taxes paid.

From the fourth quarter (Q4) of 2017 to Q3 of 2018 inclusive, the total number of EU nationals working in the UK fell by 132,000 (5.5%)

Table 6.4 Migrant workers across UK tourism and hospitality sub-sectors, 2011 and 2016

Industries	2011			2016			Increase in migrant worker numbers 2011–2016
	Number of migrant workers	% of migrant workers in sector	% of migrant workers from other EU countries	Number of migrant workers	% of migrant workers in sector	% of migrant workers from other EU countries	
Hotels and similar accommodation	65,794	26	23	99,645	35	58	33,851
Holiday and other short - stay accommodation	3,823	7	1	5,890	7	79	2,067
Camp sites and other accommodation	2,091	11	1	3,326	16	76	1,235
Restaurants	250,659	37	52	298,336	35	40	47,677
Food and service management	30,644	19	7	45,385	22	36	14,741
Pubs, bars and nightclubs	22,482	8	8	19,533	7	45	(–2,949)
Organisation of conventions and trade shows	5,704	31	2	2,507	10	30	(–3,197)
Tourist services	1,700	8	1	4,874	14	16	3,174
Museum and cultural attractions	4,520	9	1	9,396	17	45	4,876
Visitor attractions	17,823	7	4	25,425	9	68	7,602
TOTAL	405,239	22	38	514,317	24	45	109,078

Source: People 1st., 2017a: 6.

Table 6.5 (Non-UK) EU employment in tourism-related UK sectors, 2018

Sector	Key statistics	Detail
Agriculture and Horticulture	The horticulture sector alone required 60,000 seasonal workers for harvesting fruit, vegetable and flower crops, 75% of whom hitherto had come from Bulgaria and Romania	Migrant workers move within the UK as the season progresses for different crops. In 2017 farms recorded a 12.5% shortage of seasonal workers. Available automation could not meet the needs of planting, picking, grading and packing delicate crops.
Creative industries	131,000 EU nationals comprise almost 7% of the sector's workforce	Mobility of workers to and from Europe is critical for the sector.
Food and drink	EU migrant workers make up 30% of the sector's workforce	A number of roles, not skilled in the definitional sense, require delicacy and are not easily automated.
Public transport	Up to 20% of the railway industry's workforce are EU nationals	EU workers also help fill critical gaps in such diverse areas as bus driving and aviation engineering.
Retail	170,000 EU nationals comprise almost 6% of the sector's direct workforce	Retailers are especially dependent upon EU workers in such supply chains as food and drink, logistics and warehousing.

Sources: CBI, 2018; Lang *et al.*, 2018; NFU, 2017.

to 2.25m, the largest annual absolute reduction since comparable records began in 1997 (ONS, 2018). But this reduction was not to last, such that over the two-year period from 2017 Q1 to 2019 Q1 EU nationals working in the UK increased in number by 3.2%, from 2.31m to 2.38m. However, non-EU nationals in employment increased by 8.0% from 1.22m to 1.32m (ONS, 2019a: 23).

As part of the transposition of the *acquis* of EU law into UK law (Chapter 2), the UK government confirmed that it would 'entrench all existing workers' rights in British law' (Ford, 2016: 398), although there was some debate as to what this would mean in practice (Chapter 10).

An indication of the significance of EU workers in wider tourism-related sectors presented in Table 6.5 emphasises the multiple, often fragmented influences exerted on tourism-related sectors.

The obverse were those countries of Central and Eastern Europe, such as Romania (Bâc, 2015), that had been suffering skilled labour shortages as a result of the out-migration of young workers following accession to the EU. They could now benefit from the return of some of those migrants who had been contributing to economic growth in the UK, and who would take home valuable skills and experience.

6.1.2 Employment and migration controls

With the likely loss of free movement for EU nationals to the UK, the migration system needed to be revised as most of the people required

Box 6.1 (Pre-*Brexit*) Home Office points-based system for UK visa applicants

Tier 1: General (extension only), entrepreneurs, investors, post-study workers (coming to an end), exceptional talent, graduate entrepreneur: this has been the visa of choice for professionals and employers because of the flexibility provided.

Tier 1: Investor programme: 'golden' visas sold to wealthy foreign citizens (allegedly for around £2m). Although planned to be suspended from December 2018, the scheme continued.[2]

Tier 2: General, ministers of religion, intra-company transfers and sports persons: must have a UK employer who wishes to issue them with a certificate of sponsorship.

Tier 3: Low-skilled workers: this category had been suspended.

Tier 4: General, child student, student visitor, child visitor, prospective student wishing to study in the UK: must have a confirmation of acceptance for studies from their educational institution.

Tier 5: Youth mobility scheme and temporary workers: must hold a certificate of sponsorship.

Source: House of Commons Library, 2018; Pegg and Grierson, 2018.
Note: This system was subject to likely early revision.
See also Box 12.1.

to fill the vacancies in the tourism and hospitality sectors would not qualify for entry under either Tier 1 or Tier 2 (Box 6.1). Short-term options such as using seasonal work permits to fill vacancies would be only partly appropriate as major destinations operating year-round experienced limited seasonal impact. Also, seasonal permits would not support the sector's commitment to develop skilled staff (Tourism Alliance, 2017: 8–9).

The tourism and hospitality sector required long-term employees with a wide range of 'soft skills' in such areas as customer service, gastronomy and languages. The UK government needed to adapt its immigration scheme so that these factors were taken into consideration when evaluating what constituted a 'skilled' job. It therefore needed to collaborate with the tourism and hospitality sectors to develop and implement a plan for filling staff shortages to enable growth in a post-*Brexit* environment.

A report by the UK government's Migration Advisory Committee (2018) which provided the basis for a subsequent government policy paper on immigration (UK Government, 2018e), ignored the requirements of the hospitality (and haulage and house-building sectors) by proposing that the existing Tier 2 skilled workers scheme, which applied to people from outside the EU/EEA for skilled jobs earning more than £30,000 a year (a figure subsequently subject to a 12 month consultation period), could act as a template for post-*Brexit* migration policy, removing the cap of 20,700 on this source.

This apparent restriction in favour of 'higher skilled' workers being allowed visas, expressed no preferential treatment for EU citizens and appeared to ignore the needs of those sectors dependent upon less skilled migrant workers, who would be able to work in the UK for just 12 months. The Institute for Public Policy Research (IPPR) argued that such a policy would mean that about 75% of the UK's EU workforce would not be eligible for skilled status, highlighting, for example, that an estimated 94.5% of EU migrants working in north-east England would not qualify (Griffith & Morris, 2017).

For UK citizens working in the EU, the Posted Workers Directive had supported the operational delivery of travel products and services by enabling travel businesses to temporarily place workers in other EU countries without the need to register individuals in each territory for the purposes of taxation or social security. Many UK tour operators, especially SMEs and those without registered businesses in other EU member states, had relied on this regulation to simplify their operations across the EU.

The removal or restriction of such employment rights for English-speaking holiday company representatives located in popular hotels and resorts would represent a substantial burden for UK travel companies, constraining their ability to compete while significantly increasing operating costs and complexity. These employment rights needed to be retained in negotiations on a future trading relationship.

In the event of a 'no deal' exit, 25,000 seasonal jobs held by UK citizens in European ski resorts and summer activity centres would be at risk. This sub-sector was claimed to generate more than £16bn for the UK economy and £1bn in UK taxes. A consequent *Brexodus* of UK staff from alpine resorts could potentially raise the cost of a British package holiday by more than 30% (O'Carroll, 2018b).

6.2 Education and Training

In relation to tourism and travel, the UK education and training sector provides at least four important functions: (i) providing both the 'soft' and 'hard' employment skills required for tourism and travel occupations; (ii) stimulating tourism and travel by attracting overseas students (and staff); (iii) being enmeshed in EU and other student and staff exchange and mobility programmes; and (iv) generating large numbers of education and research-based conferences (as well as external examination and quality assessment travel and hospitality) both within and outside the UK. As such, there are three major mobility dimensions relating to the impacts on the education sector of UK withdrawal from the EU.

6.2.1 Potential loss of student mobility, its educational and experiential benefits

The Erasmus/Erasmus+ European student exchange programmes have broadened the horizons of more than 9 million students within the EU – including more than 200,000 from the UK – since their inception more than 30 years. They have exposed students to different perspectives and ideas, such that:

> ...the unquantifiable beauty of Erasmus is its ability to bring together young Europeans with similar interests and temperaments who would normally be segregated by accident of birth and national border ... Erasmus has formalised and opened up a trend that used to be the pre-serve of the privileged few... It's no wonder the present government has so little to say on the subject. None at the top table are young enough to have any first-hand experience of it, and are at odds with the cosmopolitan mindset of the Erasmus generation. (Arscott, 2017: 27)

A single UK government 'guarantee' was that university research funding hitherto underpinned by the EU within the Multiannual Financial Framework (MFF) 2014–2020 – disbursing €14.7bn in grants for exchange and collaboration over the seven years – would be subsequently funded by the UK government to cover payment of awards to UK applicants for all successful Erasmus+ bids submitted before 29 March 2019 (UK Government, 2018b).

There was no commitment to offer support beyond the conclusion of 2014–2020 projects. In the event of a no-deal *Brexit*, the European Parliament voted to guarantee Erasmus+ funding both for UK students already studying in Europe and for EU students in the UK.

6.2.2 UK citizens' linguistic ability to engage both with international visitors to the UK and with the wider world

The decision in 2004 to make the availability of modern languages optional at GCSE level was perhaps the beginning of a process that has seen the closure or merger of many UK school and university language departments (Arscott, 2017). The British Council (2017) argued that, for the UK to succeed after EU withdrawal, an emphasis on language learning would need to become a national priority, otherwise the country would be poorer economically and culturally. Mandarin[3] and Arabic were among the languages the UK would need, as well as Spanish, French and German. Only a third of Britons were able to hold a conversation in another language. Dwindling numbers were choosing to study a modern foreign language either at school or university: 2018 saw a further 8% decline in the numbers of modern language students examined at 'A' level (Bennett & Woolcock, 2018).

A society that relied on other people to speak its language could render itself perpetually vulnerable and potentially isolated. Ability to speak foreign languages rated highly as an element of soft power (Chapter 4), representing empathy and respect.

The British Council's (2016) checklist submitted to the All-Party Parliamentary Group on modern languages sought four essential language-specific objectives of the *Brexit* process (Box 6.2).

Box 6.2 The British Council's *Brexit* checklist submitted to the All-Party Parliamentary Group on modern languages

1. Guarantee residency status for EU nationals already living in the UK and safeguard future recruitment of EU citizens to address the shortage of language skills:
   - an estimated 35% of modern foreign language teachers and 85% of modern foreign language assistants in UK schools were non-UK EU nationals;
   - the UK alone was not producing enough languages graduates to fill the teacher shortage – estimated at 3500.
2. Ensure access to and participation in the Erasmus+ programme (as do Norway and Switzerland):
   - employers favour graduates (in all subjects, not just linguists) who have spent a year abroad and have acquired language and intercultural skills;
   - without Erasmus+, UK graduates would be disadvantaged in a global labour market.
3. A commitment to transpose into UK legislation the rights enshrined in the 2010 European Directive on the Right to Interpretation and Translation in Criminal Proceedings.
4. A post-*Brexit* plan in education and training, business and the civil service, with specific actions to ensure the UK produces sufficient linguists to meet its future requirements as a leader in global free trade:
   - the UK's language skills deficit was estimated to cost 3.5% of GDP;
   - trade negotiations and other key functions previously carried out by the EU would require UK officials with language skills;
   - 83% of UK SMEs operated only in English, yet over half claimed language skills would help expand business opportunities and build export growth.

Source: British Council, 2016.

6.2.3 Potential loss of EU staff from UK educational institutions

British universities warned the government they risked losing talented EU staff because of the lack of clarity concerning their post-*Brexit* rights in the UK (Universities UK, 2018). The potential risk to UK universities from potential academic flight was further highlighted in a British Academy (2017) report. Almost 40,000 non-UK EU staff worked in UK universities, making up 12% of all full-time equivalent staff across the higher education sector. Regionally, the risk was particularly acute in Northern Ireland where a quarter of all academic staff – across all subjects – was from EU countries, while in the West Midlands almost half of modern languages staff was from the EU. Overall, the humanities and social sciences would be particularly hard hit, embracing six of the top 10 subjects with the highest proportions of non-UK EU staff.

6.3 International Transport

Regulations dealing with the inter-operability of transport systems across the EU include the Single European Sky (1994): a set of four regulations designed to ensure the smooth operation of flights, increase efficiency, and reduce delays within European airspace. There has also been comparable legislation in place for rail and maritime operations.

The EU's aviation industry is one of the most liberalised in the world. There is significant collaboration between EU member states to facilitate operation of an internal market for aviation, which guarantees EU airlines the right to operate point to point air routes anywhere within the EU. These rules are enshrined in Article 119 of the TEU.

Such freedoms have allowed low-cost carriers such as *easyJet* and *Ryanair* to flourish, forcing 'legacy' carriers to cut costs and fares. As a consequence, the number of routes across Europe increased by 180% and fares had fallen on average by over 40% in the decade or so since the mid-1990s (ABTA, 2017): with no little cost to the global environment.

In addition, the EU has negotiated 'Open Skies' agreements, which are effectively bilateral agreements with third countries, notably facilitating longer-haul routes, including all flights to the USA. Thus *British Airways* and *Virgin Atlantic* have, hitherto, secured easy access to America.

Given the importance of transport, and especially aviation, to the UK economy, it was logical for the UK to prioritise continued cooperation in these spheres. However, whilst trade deals could be reached with the EU to enable continued market access and regulatory compliance, by leaving the EU the UK would lose its influence in shaping future regulation (Deloitte, 2016: 9). Further, in the event of agreements not being secured, there was no international fall-back option, such as World Trade Organisation rules, that existed for the aviation sector (ABTA, 2017).

Thus the freedom for UK airlines such as *easyJet* to fly within and between EU countries was potentially threatened. Despite the EU indicating that flights should continue even in the event of a 'no-deal' UK departure, *easyJet* set up a European company, based in Vienna, obtaining an EU air operating certificate (AOC). Conversely, EU-based airlines *Ryanair* and *Wizz Air* sought and secured UK AOCs from the Civil Aviation Authority. Nonetheless, depending on the final terms of a UK-EU trade agreement, bilateral treaties would possibly still need to be negotiated, adding complexity and cost to operations.

British shipping company P&O re-registered under the Cypriot flag in order to avoid complications for its UK-EU ferry, cruise and freight services.

If the UK negotiated a similar arrangement to Norway, within the European Economic Area (EEA), then little would change; *Norwegian*, a non-EU budget airline, flies unimpeded within Europe and from the UK to the US.

In addition to aviation, around 19m journeys have been under-taken each year between the UK and the EU by rail, sea and road. Joint approaches to regulation under EU auspices, including the mutual recog-nition of qualifications and drivers' working hours, have enabled these transport links (ABTA, 2017). Indeed, North Sea ferry timetables are mostly geared to facilitate drivers' rest time requirements.

Substantial disruption was anticipated if appropriate agreements were not in place for the UK's withdrawal. UK government interim plans to employ channel port motorway approaches as stacking areas for goods vehicles delayed by new customs requirements were not auspicious. Neither was the contracting of (in one case putative) ferry companies to secure extra capacity for the movement of vital goods in anticipation of a 'no-deal' *Brexit*. Storage costs increased dramatically as stockpiling resulted in a shortage of warehouse space, not least for phar-maceuticals and food (Butler, 2019).

6.4 The Food Chain

Not only has the UK relied on significant numbers of non-UK EU nationals providing labour to service critical points along the food chain (section 6.2 above), but the fragile state of UK food security was further placed in jeopardy by the fact that the UK sourced around 30% of its food from the EU, with a further 11% from other countries via agree-ments negotiated by the EU (Lang *et al.*, 2018). The UK government acknowledged this threat by planning to suspend food regulations in the event of a no-deal *Brexit* in order to maintain the flow of food.

The UK has been experiencing a widening food trade gap: in 2017 importing food, drink and animal feed worth £46.2bn while exporting only £22bn-worth, of which whisky accounted for nearly a fifth (DEFRA, 2018a). While there were calls amongst Conservative Party *Brexiters* for the UK to pursue 'large' trade deals with America, the US Commerce Secretary insisted that abolishing EU food standards was a prerequisite for any UK-US free trade deal. US food exports to the UK had been relatively small, only 10th most important after nine EU sources led by the Netherlands (DEFRA, 2018a) (Figures 6.1–6.2).

Figure 6.1 Containers in Rotterdam docks ready for transhipment across the North Sea

Figure 6.2 Line of Dutch refrigerated lorries on the Amsterdam to Newcastle ferry

There were at least three serious potential threats to UK food security (Lang *et al.*, 2018):

- Business continuity: contracts for food supplies are typically set twelve months ahead. UK food comes via a complex logistics system run on a just-in-time basis, usually three to five days' supply. Relatively small food stocks are held in the UK's food distribution chain.
- Food safety: inspections of food took place 'at source' while the UK was within the EU. Exiting the customs union and single market would require inspections to take place at UK ports, including airports for air-freighted foods. Hitherto, port inspection of EU-derived food had entailed paperwork completion that took an average of two minutes. In the event of a 'hard' *Brexit* and the establishment of customs and food safety clearances, increasing the delay by a further two minutes per lorry would generate tailbacks extending to more than 17 miles within the first 24 hours. This would inevitably shorten the shelf-life of food and increase prices – some perishables would perish and more lorries would be needed – with obvious knock-on effects for the hospitality sector and the environment.
- Home-produced UK food supplies: although slowly declining overall since a high point in the 1980s, the potential for home production would be significantly affected by whichever post-EU framework was adopted. One study suggested that farm incomes would more than halve if the UK opened its borders to a low-cost regime; they would drop by less than half if the UK adopted a unilateral protectionist regime, and incomes would rise if a free-trade agreement was negotiated with the EU (AHDB, 2017).

Warnings that the reported stockpiling of non-perishable goods in preparation for a potential 'no-deal' *Brexit* (e.g. Wood & Kollewe, 2019) had unleashed a 'bullwhip effect', whereby fluctuations in demand are amplified as they ripple back along the supply chain, destabilising production and distribution operations, were largely unacknowledged (McKinnon & Fransoo, 2019).

6.5 Funding for Tourism Development

As hitherto the third largest EU budget contributor, of the £13bn that the UK has annually paid, £4.5bn has been returned to the UK in such disbursements as Common Agricultural Policy (CAP) payments to farmers or Objective 1 funding for poorer regions such as Wales and Cornwall. This funding has provided significant benefits for the UK tourism sector. In 2016, the CAP-funded Rural Development Programme administered through DEFRA included £20m specifically targeted toward rural tourism promotion, while rural tourism businesses could access £138m of LEADER programme funds and a further £178m from the EAFRD (European Agricultural Fund for Rural Development) Growth Programme.[4] The £3.1bn available for environmental schemes through the Rural Development Programme helped to enhance the rural environment and render it potentially more attractive to visitors (Tourism Alliance, 2016: 6).

While this funding would end when the UK left the EU, it was uncertain as to how far the UK government would be willing to replace it.

The 365m trips made to rural destinations each year generate £18.6bn and a claimed 340,000 full-time jobs (Tourism Alliance, 2017: 6). As such, tourism and recreation generate more revenue and employment for rural areas than does agriculture. Yet the value of tourism to the rural economy remains poorly understood. Funding available for tourism development through CAP was only a small fraction of the €27.7bn in payments received between 2014 and 2020.

The UK needed to develop and implement a replacement scheme for CAP in order to support the agricultural sector and maintain the UK's rural landscapes. Leaving the EU provided an opportunity to reassess the incentive structures that had underpinned CAP payments, and for payments to acknowledge the strong and growing importance of tourism and recreation to rural economy and society. A scheme was required that recognised the linkages between agriculture, the environment and sustaining landscape quality in order to develop a more holistic approach to rural-based tourism and recreation (Tourism Alliance, 2017: 6).

Many councils identified European funding as a source of existing or potential income for local museums (Local Government Association, 2016, 2018). The UK's leaving the EU would effectively remove funding sources for museums working internationally (Andrew, 2016), such as Creative Europe (https://ec.europa.eu/programmes/creative-europe/about_en), the EU programme assisting cultural cooperation with a 2014–2020 budget of €1.46bn.

The UK had been successful in accessing these funds. In 2016, UK-based organisations were involved in just over 50% of successful Culture Co-operation Project applications, rendering the UK the best networked and most involved of all EU partner countries (Andrew, 2016). This would be lost after 2020.

Pessimism was also expressed over the likely impacts of EU withdrawal on archaeology and heritage management: not simply over the loss of European funding sources, but also the likely reductions of UK-based EU personnel, as well as the uncertain legal context in which work would be undertaken. The apparent denigration of expertise arising out of the referendum's 'Leave' campaign suggested the threat of populist undertones for heritage funding in a post-*Brexit* UK (Schlanger, 2017).

6.6 Sector Taxation and Regulation

The EU's Customs Union governs tariff-free trade within the EU single market. If the UK was no longer part of this, rules relating to VAT and the Tour Operators Margins Scheme (TOMS)[5] would be in doubt. In such circumstances the UK would have the option to adopt a system similar to TOMS, not apply VAT to any travel-related services that were delivered outside the UK, or apply VAT to all outbound tourism sold by UK tour operators. While the first option might cause the least change to existing practises, it was still likely to require significant efforts to formulate and implement.

Although air passenger duty (APD) is not a European Union tax, its application could change with UK withdrawal. Under the EU provision of free movement of people, connecting passengers in the UK have been exempt from APD. But outside the EU, the UK could adopt more flexibility in applying APD to connecting passengers and also in employing different rates of APD on domestic and international flights.

Although the Scottish Government reduced air passenger duty by 50% in 2018, Tourism Alliance (2017) considered it a priority for the UK to abolish APD on all domestic and international journeys departing UK airports as soon as practicable, viewing it as a tax which penalised both UK citizens and those visiting the UK. Other countries' reduction or removal of aviation taxes had contributed to growth in air transport and tourism.

On questionable environmental grounds, Tourism Alliance (2017) further argued that by removing APD in the UK increased frequencies on existing air routes would follow, while encouraging 'the introduction of new routes to the world's fastest growing economies, all of which will be vital in a post-*Brexit* environment' (Tourism Alliance, 2017: 11).

While no longer being bound by EU tax harmonisation might enable a post-*Brexit* UK to implement competitive tax policies, the position of UK tourism and travel businesses vis-à-vis their European counterparts could be adversely affected. The UK would still need to be compliant with OECD principles which play a crucial role in driving global as well as EU tax laws. Agreements with the EU would then determine whether UK businesses could benefit or face additional barriers to trade when dealing with EU countries.

In 2017, the World Economic Forum, while ranking the UK as the 5th most competitive destination overall, ranked it 140th out of 141 countries in terms of price competitiveness, mainly due to visitors facing relatively high levels of taxation (Crotti & Misrahi, 2017). Tourism Alliance (2017) therefore argued that government needed to work with the tourism and hospitality sector to conduct a review of taxation to produce a post-*Brexit* regime that was fair and stimulated growth.

The nature of the tourism-related accommodation sector had been significantly modified through the 'sharing economy' generated by such 'disruptive' cyberspace facilitators as *Airbnb*, *Wimdu*, *Arbitel* and *9Flats*, operating on the margins of regulatory compliance, not least in encouraging the (contested) pervasive inflationary presence of tourists in residential areas (Mody *et al.*, 2019). Yet the UK government had not indicated that it would intervene in the operations of such facilitators, despite measures having been taken elsewhere in Europe.

6.7 The Reliability of Mobility Data

The then UK home secretary, Theresa May, had promised that an e-system of 100% exit checks would be in place at Britain's borders by 2015. But a report by the chief inspector of borders and immigration (Bolt, 2018), found the system contained no departure records for more than 600,000 people who should have left the country in the previous two years. There was also a growing 'unmatched pot' of more than 201,000 records of departures of people for whom there was no Home Office record that they had entered the country.

Because it was not possible to be certain about a person's movements, the Home Office could not rely on the system as evidence of immigration (non-) compliance. By June 2017, significant gaps remained in collecting data on people arriving by sea, for rail journeys in and out of Britain and travel to and from Ireland within the common travel area (Chapter 9). Between April 2015 and March 2017 the Home Office collected on its database more than 607m records of departures from Britain as a result of the exit check programme. The vast majority were for UK or European nationals. However, the report established that by March 2017 there were 88,134 visa visitors – from outside Europe – whose visas had expired within the previous two years but for whom there was no record of departure.

A Home Office special investigation was launched in 2017 when it was identified that 8,474 out of 52,238 Chinese visa holders who were required to leave the UK between April 2015 and March 2016 had not been recorded as having left the country. Chinese visa holders are considered a low-risk group by the Home Office. The UK embassy in Beijing contacted Chinese tour operators and secured evidence from each traveller that had returned to China, including a scan of their re-entry stamp. The investigation concluded that the vast majority had returned home but their departure had not been captured on the UK system. This was because they had left the UK by ferry or train or via the common travel area or on a flight where data were incomplete.

The report emphasised that this work needed better coordination within the Home Office, and externally with carriers, with other potential contributors to, and users of, the data, and with common travel area partners. In particular, the Home Office needed to be more careful about presenting exit checks as a means of managing illegal migration (Bolt, 2018). This would be critical for any post-*Brexit* control regime.

6.8 Conclusion

Unsurprisingly, supply-side issues that would need to be resolved following the UK's withdrawal from the EU appeared somewhat more challenging than the government's *Tourism Action Plan* (DCMS, 2016) suggested. The following chapter addresses the likely major demand-side issues.

Notes

(1) Forecasts produced for the Local Government Association estimated that by 2024 a growing skills gap would result in a shortage of 4.2m skilled people in the UK, equating to a loss of £90bn in economic output (Local Government Association, 2017b, 2018).

(2) In 2015 Transparency International UK claimed that 3,000 'golden' visas had been issued without checks on the source of funds employed to acquire them.

(3) In 2018 Mandarin overtook German as the modern foreign language with the third highest number of students examined at 'A' and 'AS' level (in England, Wales and Northern Ireland) (Bennett & Woolcock, 2018; JCQ, 2018). See also Chapter 13.

(4) For the 2014–2020 programme period, the EAFRD's three main objectives related to agricultural competitiveness, natural resource sustainability, and balanced territorial development of rural economies and communities (http://ec.europa.eu/regional_policy/en/policy/what/glossary/e/european-agricultural-fund-for-rural-development).

(5) TOMS is a VAT scheme aimed at simplifying the rules for suppliers of designated travel services. It applies not just to tour operators, but often also to travel agents, event organisers and other businesses selling travel. TOMS imposes a VAT charge on the gross margin made by the operator rather than the usual 20% applied to gross sales, and VAT on the costs of travel supplied, therefore, cannot be recovered (Gibson, 2014).

7 Demand-side Issues

This chapter briefly examines some of the actual and potential impacts of the UK's departure from the EU on domestic, outbound and inbound visitor dynamics.

7.1 Immediate Impact: Weaker Sterling

Following the EU referendum, sterling's value fell 10% against the euro from 1.31 to 1.15. A weaker pound was therefore likely to encourage domestic trips (austerity constraints notwithstanding) and European inbound tourism, but discourage UK outbound tourism as overseas travel became more expensive (VisitBritain, 2016a). Sterling's decrease in value against the US dollar, also attracted longer haul inbound visitors to the UK. Holiday search engine *CheapFlights*, reported that the days following the referendum saw flight searches to the UK from the US more than double. Online travel agency *Opodo* reported a 38% increase in searches for flights to the UK from US markets, 47% from Canada, 31% from Europe, and a 20% increase from Asian territories including a 61% rise from China (Reckless Agency, 2017).

The immediate post-referendum perception of bargains was explicit, with UK prices for North American visitors 10–15% lower than pre-referendum levels, and 30% lower than two years previously. And because fewer British and European travellers were flying to the US, airfares were already historically low. The day after the referendum vote fares to London from the US on Virgin Atlantic and other airlines fell to $500 return. At least until the UK actually left the EU, London would continue to be seen as a major travel hub for onward journeys into Europe.

On the other hand, there were expressed concerns in the US travel trade that, after the UK's EU withdrawal, hotel rates, tour and car driver rates, food costs and other travel expenses would increase, especially if non-UK citizens lost their service industry jobs and the cost of labour rose. All the more reason to make plans to travel to the UK before that time, was the recommendation. As a consequence, US (and total) visitor numbers peaked in 2017 (Tables 7.7 and 7.8, pp. 108–109).

7.2 Domestic Tourism Effects

While the overall trend in both number of domestic tourist trips and nights spent in *Britain* between 2011 and 2017 saw a decrease of 0.8%

Table 7.1 GB domestic overnight tourism, 2016–2017

	Trips (in millions)			Bed-nights (in millions)			Expenditure (£bn)		
	2016	2017	% growth	2016	2017	% growth	2016	2017	% growth
GB	119.5	120.7	1.0	359.6	369.4	2.7	23.08	23.68	2.6
England	99.3	100.6	1.3	287.7	299.4	4.1	18.5	19.05	2.7
Scotland	11.5	11.7	1.7	38.9	39.1	0.5	2.9	3.01	3.5
Wales	9.3	9.0	–3.2	33.0	31.0	–6.1	1.7	1.63	–5.9

Source: Kantar TNS, 2018a: 19.

per annum, for 2016–17 they experienced an increase of 1.0% and 2.7% respectively, with spend increasing by 2.6% to £23.68bn (Table 7.1). However, the volume of domestic overnight business trips in *Britain* declined in 2017 by 1.8% with expenditure down by 4.6% (Kantar TNS, 2018b: 19–20), reflecting a pre-*Brexit* effect of business uncertainty.

Tourism Alliance (2017: 8–9) had estimated that domestic tourism would increase by more than 4% in 2017 in response to the increased cost of overseas holidays for UK travellers. The lower than predicted growth could be attributed to continuing austerity and the uncertainty arising from the slow pace of EU withdrawal negotiations and their possible outcomes.

7.3 Outbound Tourism

Ranked as the world's fourth highest outbound tourism spender in 2016 (UNWTO, 2017a: 14), the UK outbound tourism sector directly contributes over £26bn per annum to the British economy and provides employment for over 400,000 people in the UK. The sector argues that this reflects the UK's position as one of the world's key hubs for international travel and represents a strong argument for expanding the UK's international transport network which, in turn, could facilitate wider trade in a post-EU 'global' environment (Tourism Alliance, 2017: 4–5). Outbound tourism's role in soft power projection and the generation of a positive national image is less certain (Chapter 4).

Despite predictions that sterling's devaluation following the EU referendum vote would exert a constraining effect on outbound tourism, Table 7.2 suggests that the annual increase in outbound numbers for 2016 was actually higher than average at 7.7%, compared to the 2013–17 average of 5.9%. However, with continuing *Brexit* uncertainty the rate of growth for 2017 was noticeably lower (2.8%), while in 2018 an absolute decline set in for total outbound numbers (–1.4%), for those to EU (–1.9%) and to Commonwealth destinations (–3.8%: see also Chapter 12).

Table 7.2 UK outbound visitor numbers, 2014–2018

UK outbound visitor numbers (m)	2014	2015	2016	2017	2018	Annual average % growth 2014–18
Total	60.082	65.720	70.815	72.772	71.733	4.5
To EU	43.834	48.113	52.954	54.680	53.658	5.2
To Commonwealth	5.285	5.671	6.565	6.523	6.275	4.3
% of total UK outbound						
To EU	73.0	73.2	74.8	75.1	74.6	–
To Commonwealth	8.8	8.6	9.3	9.0	8.7	–

Sources: ONS, 2019b: Table 3.10; author's additional calculations.

Outbound visitor spend continued to increase (Table 7.3). With the lower purchasing power of the pound one might have expected a growth in outbound spend proportionately greater than the (until 2018) increases in visitor numbers noted above. In fact, with an average 2013–17 growth in spend of 6.8%, 2016 revealed an increase of 12.1% growth – indeed greater than that of visitor numbers – while 2017 returned a modest 2.4% increase, a figure lower than that for visitor numbers. By contrast, despite an overall decline in outbound visitor numbers for 2018, spending continued to grow, albeit by just 1.3% (ONS, 2019b: Table 3.11). While spending in EU destinations continued to increase both absolutely and proportionately, it declined in Commonwealth destinations.

Table 7.3 nonetheless reveals the notable difference between the spend levels of the three-quarters of all UK outbound visitors travelling to the EU and the 9% visiting the Commonwealth, whereby per

Table 7.3 UK outbound visitor spend, 2014–2018

UK outbound visitor spend (£bn)	2014	2015	2016	2017	2018	Annual average % growth 2014–18
Total	35.537	39.028	43.771	44.840	45.435	6.3
In EU	19.759	22.158	25.398	26.741	27.506	8.6
In Commonwealth	4.985	5.040	6.208	6.388	5.883	4.5
% of total UK outbound spend						
In EU	55.6	56.8	58.0	59.6	60.5	–
In Commonwealth	14.0	12.9	14.2	14.2	12.9	–
Outbound visitor spend per capita (£)						
Total	591	594	618	616	633	1.8
In EU	450	460	479	489	512	3.4
In Commonwealth	943	889	946	979	938	–0.1

Sources: ONS, 2019b: Tables 3.11, 3.13; author's additional calculations.

capita spend in/to the EU – with shorter, less expensive journeys and the availability of relatively inexpensive packages – was just over half of the average spend in/to Commonwealth countries, although the gap appeared to be closing (Chapter 12).

With thirteen of the UK outbound top 20 destinations being located within the EU, Table 7.4 highlights a 2018 (temporary?) brake on the apparent inexorable growth of UK visits to Spain (although spend continued to increase: Table 7.5, p. 102). Specific data from the Costa Blanca hotel association Hosbec, for example, suggested an 8% reduction in UK overnight stays in Benidorm for 2018, albeit more than compensated for by an increase in Spanish domestic tourists (Murray, 2019).

Overall, UK outbound growth figures for Poland and Romania may reflect VFR-related visits, as will those for India. Across both outgoing and incoming visitor data the 'pre-exit years' expressed a degree of volatility, partly reflecting the uncertainty of *Brexit*, a variability in sterling's

Table 7.4 Top 20 UK outbound destinations by visitor numbers (000s visitors), 2014–2018

Destination	2014	2015	2016	2017	2018	Annual average % growth 2014–18
1. Spain	12,246	12,988	14,676	15,872	15,618	6.3
2. France	8,784	8,849	8,542	8,862	8,556	–0.7
3. Italy	2,948	3,533	4,089	4,160	4,325	10.1
4. USA	3,257	3,503	3,597	3,401	3,472	1.6
5. Republic of Ireland	3,095	3,504	3,721	3,422	3,218	1.0
6. Portugal	2,192	2,602	2,806	2,876	2,818	6.5
7. Germany	2,323	2,592	2,732	2,909	2,813	4.9
8. Netherlands	2,111	2,548	2,761	2,660	2,716	6.5
9. Poland	1,693	2,033	2,424	2,672	2,472	9.9
10. Greece	1,933	2,314	2,480	2,383	2,468	6.3
11. Belgium	1,740	1,637	1,533	1,617	1,525	–3.2
12. Turkey	1,406	1,526	1,057	1,228	1,425	0.3
13. Romania	340	478	775	1,009	1,006	31.2
14. Switzerland	848	926	1,007	955	995	4.1
15. India	884	931	990	1,006	980	2.6
16. UAE	727	786	922	1,016	836	3.6
17. Cyprus	606	650	840	901	745	5.3
18. Mexico	434	529	502	539	715	13.3
19. Malta	538	501	651	519	659	5.2
20. China (including Hong Kong)	477	537	601	569	652	9.2

Source: ONS, 2019b: Table 3.10.

Table 7.5 Top 20 UK outbound destinations by visitor spend (£m), 2014–2018

Destination	2014	2015	2016	2017	2018	Annual average % growth 2014–18
1. Spain	6,143	6,474	7,943	8,683	8.908	9.7
2. USA	3,908	4,541	4,787	4,279	4,304	2.4
3. France	3,512	3,686	3,651	3,984	4,285	5.1
4. Italy	1,681	2,034	2,360	2,378	2,609	11.6
5. Greece	1,149	1,401	1,622	1,664	1,778	11.5
6. Portugal	1,167	1,628	1,576	1,552	1,533	7.0
7. Germany	852	1,051	1,044	1,084	1138	7.5
8. Netherlands	707	815	969	974	1038	10.1
9. Republic of Ireland	1,078	1,015	1,212	993	981	−2.3
10. Poland	595	728	813	950	908	11.2
11. UAE	812	764	910	1,105	885	2.2
12. Turkey	827	878	593	712	859	0.9
13. India	774	801	808	821	702	−2.4
14. Mexico	394	472	495	593	646	13.1
15. Romania	192	243	403	547	644	35.3
16. South Africa	313	361	385	445	604	17.9
17. Thailand	591	598	767	596	596	0.2
18. Australia	815	658	842	820	575	−8.3
19. Switzerland	428	477	550	502	563	7.1
20. Canada	379	384	536	647	530	8.8

Source: ONS, 2019b: Table 3.11.

value, persisting austerity and the impacts of terrorist attacks. Some outbound destinations showed a notable increase in visits for 2016 then a decrease for the following year (USA, Switzerland and Ireland, with the latter continuing to decline: see Chapter 9). Others experienced a drop in 2016 followed by significant gains in 2017, notably Turkey, France and Belgium, albeit with the latter two destinations expressing a decline again in 2018. As the UK's almost nearest neighbours, both countries were the only UK outbound destinations in the top 20 to show an overall decline in visitor numbers for 2014–18.

As sterling had lost 20–23% of its value between the 2016 referendum and the beginning of 2019, destinations outside the eurozone such as Mexico, South Africa and Turkey were being favoured by UK sun worshippers. On the other hand, hitherto popular UK destinations such as Egypt and Tunisia had been subject to terrorist attacks and/or political instability in recent years, with resultant visitor uncertainty in the short- to medium-term (Avraham, 2016; Lanouar & Goaied, 2019).

Almost paradoxically, but partly reflecting sterling's loss of value, Table 7.5 reveals UK outbound double digit average spending growth in Italy, Greece and the Netherlands for the 2014–18 period.

A relatively safe destination outside the eurozone was Bulgaria: traditionally lower cost and more amenable to all-inclusive arrangements whereby spending on food, transport and accommodation could be secured against further sterling depreciation. During the first half of 2019, ABTA was reporting an increase of almost a third in UK bookings to the Balkan country compared to the previous year (Hickey, 2019).

Even if, following withdrawal from the EU, there were no changes to travel requirements within Europe in terms of freedom of movement, potential UK outbound tourists' travel behaviour could be influenced by such issues as access to health care, travel protection and further currency fluctuations. Table 7.6 attempts to summarise the main *Brexit*-related factors impacting on UK outbound travel decisions (see also Chapters 3 and 5).

7.3.1 Enhanced border controls

While, within the EU, a UK passport normally ensured minimal formalities, for some time (especially after the 1985 Schengen agreement), airlines had been obliged to collect 'advance passenger information' on behalf of national governments for security purposes, although the process had usually taken a light touch approach. And, of course, UK passport holders were able to use the 'fast-track' lanes for EU/EEA citizens.

Delays at borders through enhanced passport and customs checks had already been experienced following the introduction of exit checks by Schengen zone countries (European Commission, 2018j): a taste of experiences to come.

Being from a 'third country', UK passport holders would need to apply online for a three-year 'travel authorisation' to visit an EU member state from 2021 under the European Commission's European Travel Information and Authorisation System (ETIAS). This was adopted in the wake of the 'migrant crisis' and security concerns over terrorism. It would cover visits of less than three months, cost €7 (for 18–70 year-olds), and would largely emulate the United States' ESTA scheme set up after 9/11 to evaluate prospective visitors. For this, travellers would need to submit various personal details in an exercise that the Commission claimed would take no more than 10 minutes (European Commission, 2018i).

Full visa requirements for UK citizens would be waived (otherwise costing €60) provided the UK treated EU visitors similarly. The UK travel sector complained that these requirements would clearly increase the cost as well as complexity of holidays and business trips to the EU (Calder, 2017).

Table 7.6 Likely *Brexit*-related factors impacting on UK outbound travel decisions

Factor	Impact	Timing
Currency rates: £:€	More expensive holidays in the EU (1): accommodation, transport, entry fees, motorway tolls.	Post-referendum
Currency rates: £:US$	More expensive holidays in USA, Dubai, China etc. More expensive travel generally as fuel and aircraft are priced in US$.	Post-referendum
Car insurance for the EU	Increase costs likely (2). In the event of a no-deal exit, UK motorists would need to carry an international certificate of insurance ('green card') as proof of third-party cover.	Post-*Brexit*
Duty-/tax-free purchases	Pre-1999 limits re-introduced for travel to/from Europe (3). Sales opportunities for airlines and cross-Channel ferry operators.	Post-*Brexit*
Passport and customs control (see also text below)	Tougher passport control, more scrutiny from customs officials. The 'blue channel' would probably cease to exist on UK soil. The UK's 'third country' status would require travellers to purchase online a 3-year travel authorisation (ETIAS). The Schengen Border Code would demand at least 3 months' passport validity and a stay of not more than 90 days. For travel outside of the EU, the UK would be able to seek new bilateral visa agreements with non-EU countries.	Post-*Brexit*
Mobile phone roaming charges	Surcharge-free roaming, abolished within the EU/EEA in April 2017, could be re-introduced for a UK outside EU roaming regulations. Some mobile operators – *3, ee, O2* and *Vodafone* – covering 85% of UK mobile subscribers, claimed to have no plans to re-introduce surcharges. The UK government intended to retain a financial limit on mobile data usage while abroad, set at £45 monthly (4).	Post-*Brexit*
Package Travel Directive (offers consumer protection in cases of insolvency or where there is a failure to perform contracted services)	Likely to be transposed to UK law although some negotiations were required to ensure reciprocal arrangements with the EU. The existing UK ATOL scheme notionally operated to standards above the EU Directive. Over 16m travel arrangements protected by EU law were sold in the UK in 2015.	Post-*Brexit*
Health insurance implications (the European Health Insurance Card: EHIC)	Before joining the then EEC, the UK had reciprocal health agreements with many European nations, and maintained bilateral agreements with 16 other countries. If transposition and/or reciprocal arrangements for entitlement to public health care on the same basis as local people in EU countries were not concluded, the need for travel insurance would increase and premiums would likely rise. There were 26.6m EHICs in circulation within the UK: 215,608 uses were recorded in 2015.	Post-*Brexit*

Consumer Rights Directive	This underpins many important areas of consumer protection across the EU, such as clear pricing rules, information requirements and a 14-day consumer right of withdrawal for purchases of goods and services. It was incorporated into UK law through the Consumer Rights Act 2015	No change
Passenger Rights (especially EU regulation 261/2004 for compensation in cases of denied boarding or significant delays for air travel; also EU1177 for sea ferries and EU1371 for rail travel)	With UK courts often finding in favour of consumers, transposition to UK law would be in the government's interest. Notionally, EU airlines flying into the UK would still be governed by them.	Post-*Brexit*

(1) The Package Travel Regulations allow tour operators to impose surcharges when the cost of a package holiday rises after booking because of currency fluctuations or increased fuel costs. The company must absorb the first 2% of any increase; if the surcharge rises above 10% the traveller has the right to cancel.
(2) EU legislation means any car insured in one EU country is automatically insured to the minimum legal level in any other EU country. It was assumed that UK insurers would continue to extend comprehensive cover for short European trips at a minimal cost.
(3) Alcohol limits: one litre of spirits, four litres of wine and 16 litres of beer, and for tobacco 200 cigarettes; with a £390 limit on 'other goods'.
(4) Consumers and businesses in border areas were alerted to 'inadvertent roaming', when a mobile signal in a border region is stronger from the country across the border. This would be 'a commercial question for the mobile operators' (UK Government, 2018c: 3).
Sources: ABTA, 2017; Calder, 2017; Deloitte, 2016; European Commission, 2018i, 2018l; McKee & McKee, 2018; Pappas, 2019; Pickett, 2016; Reckless Agency, 2017; Travelsupermarket.com, 2017; UK Government, 2018c, 2018f.

7.3.2 Increased costs

Both in the shorter term, notably with the fall in the value of sterling, and in the longer term following EU withdrawal, travel was becoming more expensive for UK outbound travellers than might otherwise have been expected. Depending on post-*Brexit* trade agreements, there was the danger that new taxes and levies on travel could be introduced (Chapter 6), with travel businesses forced to raise prices in order to recoup costs absorbed during the transition period, such as increased supplier contracting costs owing to the position of sterling (Deloitte, 2016: 14). Car insurance would likely rise.

7.3.3 Threat to air travel

As noted in Chapter 6, the legal basis and viability of low-cost airlines operating between the UK and the EU required clarification if UK travellers were to continue to enjoy the benefits of relatively cheap and safe air travel to destinations across Europe.

According to the European Commission (2018k), EU law on passenger rights after the UK's EU withdrawal would only apply to UK passengers departing the UK if flying to an EU27 destination in an

EU27 carrier. Hence the requirement for UK-based air carriers such as *easyJet* to establish an official organisational presence within an EU27 member state since otherwise they would no longer have access to designation/traffic rights until new bilateral agreements between the EU member states and the UK as a third country were established (European Commission, 2018k).

7.3.4 Health care for travellers

Hitherto, UK residents entitled to NHS care could receive treatment for illness or injury occurring while visiting an EU/EEA country. This entitlement had been demonstrated by means of the European Health Insurance Card (EHIC), issued by the authority responsible for a traveller's health care at home (the NHS in the case of UK travellers). These arrangements were instituted to facilitate the free movement of people, acknowledging that any obstacles to obtaining health care when abroad would act as a barrier to such movement (OJEU, 2011).

After the end of the transitional period, assuming that the EHIC would no longer be available to UK residents (Fahy *et al.*, 2017), the UK could revert to the kind of bilateral agreements that preceded its EU position. If this was to be the case, UK tourists would find accessing entitlements complicated and time-consuming, and probably limited in scope (McKee & McKee, 2018). More people would need to buy private medical insurance.

In 2015/16 the UK received claims of £130.6m from other EU/EEA countries in respect of its residents requiring treatment, including £36.8m from Spain and £33.6m from France (McKee & McKee, 2018). If this was no longer covered by the NHS, it would need to be absorbed by insurers and passed on in premiums. While the additional sums would likely be relatively small for young people without pre-existing illness, they could prove prohibitive for older travellers, especially those with longer-term health conditions. McKee and McKee (2018: 136) calculated that in 2016 there were 3.9m UK travellers to other EU countries aged 65 or over, an increase of 268% since 1998. The authors warned that older people and those with chronic conditions could be discouraged by high insurance costs from purchasing any insurance at all. Or, of course, they could be discouraged from travelling outside of the UK.

Ironically, in recent years increasing numbers of UK citizens had been travelling to other EU countries (and beyond) to receive medical treatment, reportedly because of scarcer resources and unreasonable waiting times for NHS health care facilities (Germain, 2018). Of four types of medical tourists – temporary visitors requiring medical care, retirees requiring care whilst resident in another country, those travelling in their own initial to receive treatment, and those referred abroad to receive specialised care – increasing numbers of the latter two categories should become the object of post-*Brexit* UK-EU bilateral 'trade' negotiations. Otherwise, Germain (2018) argued, unregulated medical tourism would increase with potentially deleterious effects on access to domestic care as a result of wealthy patients enjoying greatest mobility, choice and purchasing power of (private) medical facilities.

7.3.5 Mutual repulsion?

The UK's withdrawal from the EU, particularly if not smooth, had the potential to set up a mutual repulsion effect between UK and EU tourists and destinations. While it was difficult to foresee UK tourists abandoning the Spanish *costas* in significant numbers, for some EU citizens, particularly if the UK government failed to satisfactorily resolve the issue of resident citizens' rights, the UK could generate negative imagery and a repulsion effect reducing the number of visits and exacerbating the UK's diminished soft power status.

Would we see a variation of the 'cultural proximity' (Kastenholz, 2010), 'self-congruity' (Beerli *et al.*, 2007), and self-selecting segmentation effects on travellers from different source areas, whereby, for example, 'Remain' voting areas such as London, Scotland and Northern Ireland would be deliberately favoured by certain European travellers, generating spatially marked patterns of (reciprocal) travel? The potential domino effects of a Scottish independence vote and subsequent (re-) accession to the EU offer further speculative variations for this hypothesis.

7.4 Inbound Tourism

7.4.1 Global Britain

As a 'global destination', the UK recorded a 4.3% growth in international visitor arrivals (ONS, 2019b: Table 2.10) and an 8.7% increase in visitor spend in 2017, despite terrorist attacks in London and Manchester (VisitBritain, 2018b) (Table 7.7). However, as travel destinations, the EU28 countries as a whole recorded an 8% growth in arrivals over the same period. The UNWTO (2018: 3) noted that the fall in the value of sterling continued to contribute to the UK's apparently good results, implying that they masked an otherwise (potentially) declining growth trend. And indeed, the following year burst the optimistic bubble in revealing a downturn of 3.3% for visitor numbers (ONS, 2019b: Table 2.10) – the first decline since 2010 – and a 6.6% reduction in visitor spend (ONS, 2019b: Table 2.11) (see Tables 7.7 and 7.9 below).

As a post-referendum phenomenon, what had those 2017 inflows symbolised? No doubt the possibly perceived temporary nature of sterling's reduced value encouraged some visits that might otherwise have

Table 7.7 Inbound visitor numbers to the UK, 2014–2018

Inbound visitor numbers (m)	2014	2015	2016	2017	2018	Annual average % growth 2014-18
Total	34.377	36.115	37.609	39.214	37.905	2.5
EU	23.009	24.213	25.486	25.586	24.795	1.9
Commonwealth	3.072	3.244	3.386	3.705	3.499	3.8
China (including Hong Kong)	0.392	0.529	0.525	0.636	0.715	20.6
% of total inbound visitors to UK						–
EU	66.9	67.0	67.8	65.2	65.4	–
Commonwealth	8.9	9.0	9.0	9.5	9.2	–
China (including Hong Kong)	1.1	1.5	1.4	1.6	1.9	–

Sources: ONS, 2019b: Table 2.10; author's additional calculations.

been put off to later years. Others may have been stimulated by logistical or other (perceived) reasons to visit to UK specifically while it was still an EU member state.

Was it continuing *Brexit* uncertainty that heralded, in 2018, the first downturn in arrivals numbers for several years? That inbound visitors from the EU27 continued to make up almost two-thirds of UK arrival numbers – albeit also reduced in 2018 – remained notable (Table 7.7). While visitors from Commonwealth countries declined in 2018, numbers from China continued to edge up (see Chapter 13).

Again, pre-withdrawal volatility could be seen in the patterns of a number of source countries (Table 7.8), with a notable reduction in visitor numbers for several in 2018. While growth from Spain, China and Romania appeared sustained, France revealed declining numbers from a 2015 peak, losing its top position to the USA for 2018. Several major EU source countries which had shown growth from 2014 to 2017, including Germany, Ireland and the Netherlands, also revealed a downturn in 2018. Incoming visitors from Poland, Italy and Portugal had notably increased in 2016 then declined in the following years.

Inbound tourism to the UK generated almost £25bn per annum for the UK economy in 2017 making it the second largest source of export earnings in the service sector after financial services, and the UK's fifth largest export earner overall (Tourism Alliance, 2017: 4; VisitBritain, 2018b: 4). Yet in 2016 the UK was only the world's sixth highest tourism revenue earner, slipping from third place in 2015 (UNWTO, 2018: 9–10), and in 2018 inbound tourism spending actually fell back by 6.6% to just under £23bn (Table 7.9). This was attributed partly to a shorter average length of stay, from 7.3 nights in 2017 to 7.0 in 2018 (ONS, 2019c: 4).

In this context of downturn, presumably borne of *Brexit* uncertainty, both the proportion of EU visitors' spending and their average

Table 7.8 Top 20 sources of visitors to the UK by visitor numbers (000s visitors), 2014–2018

Source country	2014	2015	2016	2017	2018	Average annual % growth 2014–18
1. USA	2,976	3,266	3,455	3,910	3,877	6.8
2. France	4,114	4,171	4,064	3,956	3,693	–2.7
3. Germany	3,220	3,249	3,341	3,380	3,262	0.3
4. Republic of Ireland	2,486	2,632	2,897	3,029	2,782	2.9
5. Spain	1,986	2,197	2,397	2,413	2,530	6.2
6. Netherlands	1,972	1,897	2,062	2,136	1,954	–0.2
7. Poland	1,494	1,707	1,921	1,807	1,817	5.0
8. Italy	1,757	1,794	1,990	1,779	1,808	0.7
9. Belgium	1,122	1,175	1,048	1,148	1,116	–0.1
10. Australia	1,057	1,043	982	1,092	1,003	–1.3
11. Romania	471	693	891	944	987	20.3
12. Canada	649	708	828	835	850	7.0
13. Sweden	869	850	821	831	827	–1.2
14. Switzerland	864	872	940	989	808	–1.7
15. Denmark	662	756	730	730	735	2.7
16. China (including Hong Kong)	392	529	525	636	715	8.1
17. Norway	874	771	700	712	673	–6.3
18. India	390	422	415	562	511	7.0
19. Hungary	323	328	397	415	437	7.9
20. Portugal	395	392	492	482	431	2.2

Sources: ONS, 2019b: Table 2.10; author's additional calculations.

per capita spending level increased in 2018. Commonwealth countries' proportion and average per capita remained relatively stable, while, as Chinese inbound numbers, overall spending and proportion of spending increased, per capita levels dropped slightly (Table 7.9).

While overall spend in the UK declined in 2018 for many source countries, China, Spain and Canada were notable exceptions, with China moving from sixth to fourth place in importance. Despite an absolute decline, overall spend from US visitors was still more than double of that from second placed Germany, while supporters of stronger links with the 'old dominion' Commonwealth would point to Table 7.10's indication that in recent years Australian visitors alone have provided around 10% equivalent of total EU visitor spend in the UK (see Chapter 12).

Pre-withdrawal volatility saw significant spend growth in 2016 followed by decline in 2017 and 2018 for visitors from Ireland, Italy and Switzerland, while the reverse was the case for visitors from China, Poland and Belgium (Table 7.10).

Table 7.9 Inbound visitor spend in UK, 2014–2018

Inbound visitor spend (£bn)	2014	2015	2016	2017	2018	Annual average % growth 2014–18
Total	21.849	22.072	22.543	24.507	22.897	1.2
EU	9.551	9.705	9.553	10.036	9.917	0.9
Commonwealth	3.307	3.264	3.320	3.645	3.417	0.1
China (including Hong Kong)	0.806	0.940	0.819	1.058	1.126	1.0
% of total inbound visitor spend in UK						
EU	43.7	44.0	42.4	41.0	43.3	–
Commonwealth	15.1	14.8	14.7	14.8	14.9	–
China (including Hong Kong)	3.7	4.3	3.6	4.3	4.9	–
Inbound visitor spend per capita (£)						
Total	636	611	599	625	604	–1.2
EU	410	396	386	390	400	–0.6
Commonwealth	1,076	1,006	981	981	977	–2.3
China (including Hong Kong)	2,056	1,777	1,560	1,663	1,575	–5.8

Sources: ONS, 2019b: Table 2.11; author's additional calculations.

The UNWTO (2018) predicted that international tourism would increase by 58% for international tourist arrivals to reach 1.8bn globally by 2030, with almost half that demand coming from BRIC countries. If the UK was able to regain the share of such markets it held in 2008 before biometric visas were introduced, inbound revenue could increase by a further £0.6bn per annum and consolidate the UK's position as a major destination for the world's key growth markets, presenting considerable potential opportunities for a post-*Brexit* 'Global Britain' (e.g. see Chapters 12 and 13). If, as Tourism Alliance suggested, the UK targeted growth of at least 4.2% per annum, by 2025 tourism receipts could reach £32bn per annum and by 2030 £39bn per annum (Tourism Alliance, 2017: 4). However, reduced visitor spend figures for 2018 appeared to pour a little cold water on such aspirations.

7.4.2 Student strategy

> If the government's industrial strategy is to be a success it needs a vibrant higher education sector and it is impossible to imagine that without significant and strong recruitment of international students. (Migration Advisory Committee, 2018: i)

Globally, the UK has been the second most popular higher education destination (after the USA), with around three-quarters of a million international students in higher and further education, independent

Table 7.10 Top 20 sources of visitors to the UK by visitor spend (£m), 2014–2018

Destination	2014	2015	2016	2017	2018	Annual average % growth 2014–18
1. USA	2,944	3,010	3,354	3,643	3,378	3.5
2. Germany	1,478	1,378	1,490	1,581	1,520	0.7
3. France	1,434	1,493	1,372	1,425	1,386	−0.8
4. China (including Hong Kong)	806	940	819	1,058	1,126	9.9
5. Spain	1,082	999	992	1,061	1,110	0.7
6. Australia	1,224	1,113	1,060	1,194	1,044	−3.9
7. Republic of Ireland	870	934	1,059	941	895	0.7
8. Italy	922	890	980	841	784	−4.0
9. Netherlands	701	676	714	747	716	0.5
10. Canada	473	506	634	604	676	9.3
11. UAE	437	487	566	618	616	9.0
12. India	444	433	433	454	491	2.6
13. Romania	185	215	302	299	479	26.9
14. Switzerland	488	516	792	585	460	−1.5
15. Poland	356	438	389	394	453	6.2
16. Sweden	503	510	458	451	447	−2.9
17. Belgium	342	371	305	393	399	3.9
18. Norway	548	426	411	378	394	−7.9
19. Denmark	295	356	368	359	379	6.5
20. Japan	226	214	215	250	282	5.7

Sources: ONS, 2019b: Table 2.11; author's additional calculations.

and language schools. One in five students in the UK is an international student: the highest such proportion of any destination. The economic value of this was estimated to be £10.8bn for higher education (where international student numbers had increased by 30% in the nine years to 2017), £1.2bn for English language teaching (ELT) courses, and £846m for ISC (Independent Schools Council) schools (Tourism Alliance, 2017: 14–15). In 2015 the Department for Education estimated this sector's export value at £17.6bn (Migration Advisory Committee, 2018).

Such value helps to cross-subsidise the education of domestic students and university research. It is also important to local economies with the additional spending of visits from students' families and friends.

Foreign students can play a significant role in the UK's soft power and provide the UK with the ability to gain highly skilled employees. Universities UK (2017) estimated that over 80% of former international HE students planned to develop professional links with the UK.

Most students (certainly all those on Tier 4 visas: see Box 6.1) technically fit the international definition of a migrant (i.e. those staying for more than 12 months). But this has led to students becoming unwittingly caught up in the debate about immigration, not least because there has been no cap on international student numbers. The House of Lords voted in favour of removing students from net migration statistics, and various interest groups lobbied for this to be done.

Many non-EU students have not needed to come to the UK on Tier 4 visas, preferring to arrive on short term study visas for courses up to 6 or 11 months. This has been especially the case for English language learners. More significantly, almost half of the half-million students studying at *English* UK member centres have been from EU27 countries. European Union migration rules have also assisted centres in employing qualified staff, especially those with appropriate linguistic skills.

The UK government needed to develop and implement a post-*Brexit* student tourism strategy that encouraged overseas students to study in Britain and for the best of these students to stay and further contribute to the UK's economy and society (Tourism Alliance, 2017: 14–15).

Students come to the UK from a wide range of countries, with 61 nations each being represented by at least 500 higher education students in 2016/17. International students from China are the largest group in both higher education and independent schools, and although the UK's market share of students from India has fallen sharply in recent years, the trend of students from China has remained stable (Migration Advisory Committee, 2018: 2–3) (see Chapter 13).

7.5 Regional Impacts

7.5.1 Honeypots and peripheries

Despite efforts to diffuse tourism more widely across the UK, from 2015 to 2017 the top ten UK locations that attracted the highest numbers of inbound visits remained the same while expressing enormous disparities in distribution and volume (Table 7.11), with top destination London receiving almost 10 times the number visiting second-placed Edinburgh. One 'traditional' seaside destination featured in the top 10.

If this primate city effect guided the thinking behind both the *Tourism Action Plan* and its associated *Discover England Fund*, then any spatially-based intervention in tourism growth needed to be embedded within the broader framework of regional policy, 'policy' which appeared to have become somewhat fissiparous and not a little reliant on EU funding (Chapter 5).

The wider regional distribution of incoming visitors (Table 7.12) again reflects enormous disparities between London, receiving more than half of all visits and 55% of visitor spend, with the East Midlands, North East England and Wales ('regions' with significant internal disparities) attracting the lowest numbers and percentage visitor spend.

Table 7.11 The 10 most visited UK centres by non-UK visitors, 2017

Rank	Town/city	No. of non-UK visitors ('000s)
1.	London	19,828
2.	Edinburgh	2,015
3.	Manchester	1,319
4.	Birmingham	1,117
5.	Liverpool	839
6.	Glasgow	787
7.	Bristol	602
8.	Oxford	536
9.	Cambridge	519
10.	Brighton/Hove	491

Source: VisitBritain, 2018a.

Table 7.12 Regional distribution of incoming visitors to *Britain*, 2017

Region	Numbers visiting		Overall spend		Spend per visit (£)	Average nights per visit	Average individual spend per night (£)
	millions	%	(£m)	%			
Scotland	3.2	8.2	2,300	9.4	709	8	93
Wales	1.1	2.8	369	1.5	342	6	53
England:							
West Midlands	2.3	5.9	807	3.3	348	6	57
East of England	2.4	6.1	815	3.3	339	7	51
East Midlands	1.3	3.3	444	1.8	346	8	45
London	19.8	50.5	13,500	55.1	683	6	119
North West England	3.1	7.9	1,600	6.5	508	7	75
North East England	0.56	1.4	241	1.0	434	9	48
South East England	5.3	13.5	2,100	8.6	402	7	55
South West England	2.6	6.6	1,200	4.9	466	8	58
Yorkshire and Humberside	1.3	3.3	567	2.3	423	8	55
	43.3*	(100)	24,500*	(100)			

Note: because of rounding, the final totals do not represent 100% of the column above.
Sources: VisitBritain, 2018b: 4*, 7, 8; author's additional calculations.

In terms of *English* domestic visits (see section 7.2 above), cities/large towns attracted a significant share of trip numbers and spend in 2017, but seaside resorts experienced longer stays, if not proportionately more spending. In a more even regional spread compared to patterns of overseas visitors, London did not dominate domestic tourism, but South West England attracted around a fifth of all trips, overnight stays and spend, followed by South East England (potentially accessing London), North West England, with London in fourth place (Table 7.13).

Table 7.13 Spatial distribution of domestic overnight trips in *England*, 2017

Destination type	Trips (millions)	% of trips	Nights (millions)	% of nights	Spend (£m)	% of spend
Seaside	28.0	23	107.2	29	6,006	25
City/large town	49.3	41	118.1	32	9,678	41
Small town	23.0	19	66.0	18	3,655	15
Countryside/village	21.4	18	71.8	19	3,952	17
Regions visited						
West Midlands	7.9	7	18.4	5	1,225	5
East of England	9.8	8	32.6	9	1,711	7
East Midlands	7.5	6	21.7	6	1,049	4
London	12.1	10	27.8	8	2,688	11
North West England	13.6	11	37.8	10	2,733	12
North East England	3.6	3	10.9	3	680	3
South East England	16.1	13	43.0	12	2,707	11
South West England	20.6	17	76.8	21	4,454	19
Yorkshire and Humberside	10.7	9	29.3	8	1,745	7

Source: reworked from VisitBritain, 2017a: 2.

A traditional seasonal distribution of domestic overnight trips was exhibited, with noticeable peaks in July and August, and lows in the year's first two calendar months (VisitBritain, 2017a: 3).

While the spatial disparity of visits has been greater for incoming than domestic activity, Table 7.14 reveals substantial domestic socioeconomic disparities in terms of who travels (staying overnight) and how much they spend. In *Britain* in 2017 ABs took almost four times as many trips as DEs and three times as many as C2s. The ABs also spent more than four times as much as DEs and three times more than C2s. Such figures highlight the substantial disparities both in mobility opportunities and disposable spending power. What such figures do not record, of course, are those who take no overnight trips at all (see also section 4.4.1 above).

Table 7.14 Socioeconomic distribution of domestic overnight trips in *Britain*, 2017

Social grade	Trips (millions)	% of trips	Nights (millions)	% of nights	Spend (£millions)	% of spend
AB	51.7	43	146.9	40	10,732	45
C1	36.2	30	113.0	31	7,004	30
C2	19.2	16	62.3	17	3,607	15
DE	13.5	11	47.2	13	2,341	10

Source: reworked from VisitBritain, 2017a: 5.

7.5.2 Regional policy, immobility and the *Brexit* vote

The likely differential regional impacts of *Brexit* further complicate the challenge of redesigning policy frameworks where deep-seated regional differences have persisted (Los *et al.*, 2017; UK2070 Commission, 2019). EU Structural Funds had influenced economic development in the UK for over 40 years. The UK government and devolved administrations (DAs) therefore needed to determine which domestic spatial policies to adopt from 2019 (Bachtler, 2017; Bachtler & Begg, 2017) within which tourism strategy could be embedded.

For the 2008–15 period, along with France, the UK exhibited the largest differences in regional economic performance. In 2015, the UK had the third largest regional difference in the proportion of working age population with tertiary education in the EU after Spain and Romania (Bachter & Begg, 2018). Even the 2017 *Industrial Strategy* pointed to UK regional productivity disparities being high compared to elsewhere in Europe (UK Government, 2017a).

Since the start of the financial crisis, much debate had focused on factors which explained the resilience or vulnerability of regions to sudden change/crisis (see section 4.4.1 above). Chen *et al.* (2018) found, unsurprisingly, that the UK and its regions were far more vulnerable to *Brexit*-derived trade-related risks than other EU member states and their regions. Their data suggested that the UK's exposure to *Brexit* vulnerabilities was some 4.6 times greater than that of the rest of the EU as a whole, and the UK regions were far more exposed to *Brexit* risks than regions in other EU countries, except for those in the Republic of Ireland.

One of the critical challenges for the *Tourism Action Plan* (Chapter 5) was how to link the prioritisation of 'sector deals' with local and regional economies. This partly reflected the fact that although becoming more prominent in some European regional policies, social cohesion and social justice objectives appeared not to have been prominent in recent UK regional policy practice.

Although EU regional policy spending had diminished in the UK, especially since the Central and Eastern Europe accessions, cohesion policy had exerted a strong influence in shaping UK spatial policy (Bachtler & Begg, 2017). The 2014–2020 Structural Funds programmes in the UK would continue until the end of the programme period. A UK successor 'Shared Prosperity Fund' had been mooted (Conservative Party, 2017), but its objectives, scale and allocation remained uncertain (Local Government Association, 2017a; Bachter & Begg, 2018).

There was no recognition in the *Tourism Action Plan* (nor in the *Industrial Strategy*) of the previous decade's 'hollowing out' of the capacity of local authorities and other stakeholders to implement local development.

The UK government's belated 'Stronger Towns Fund' (UK Government, 2019) aimed at 'left behind' towns in *England*, promised £1.6bn – over a third of which would be subject to competitive bid – over seven years. It was unfavourably compared to the £1bn made available for Northern Ireland to secure political support from the DUP (see Chapter 2) and the £4.2bn of Treasury expenditure on *Brexit* preparation.

The inadequate scale and politically motivated distribution of such post-*Brexit* funding was likely to exacerbate the uncertainty and apprehension felt in UK regions and localities accustomed to the multi-annual predictability of resources and support from EU regional policy (10.8bn for 2014–2020). Many of the areas that voted 'Leave', especially in northern England and Wales, would be vulnerable to continued highly constrained UK government spending if pessimistic forecasts of *Brexit* economic impacts were realised (Bachtler & Begg, 2018). Thus,

> the Brexit vote appears to have made the job of supporting and improving the conditions in the UK's more deprived areas more difficult. (Los *et al.*, 2017: 8)

There was an obvious need for a renewed focus on the relatively immobile. A wide range of literatures, not least in tourism, had perhaps placed far too much emphasis on those who were in a position to experience and benefit from various mobilities. This focus had tended to ignore a significant proportion of the population and their circumstances: research on tourism and the disadvantaged and/or disabled had developed only slowly (e.g. Agovino *et al.*, 2017; Kim & Lehto, 2012; Minnaert, 2012; Morgan *et al.*, 2015).

There was now greater justification to focus attention on the local (Lee *et al.*, 2017). Regional adaptation to globalisation and 'market integration', including the roles of tourism and hospitality, required much more differentiated, place-based, strategies (Bachtler *et al.*, 2017; Iammarino *et al.*, 2017), fostering different forms of what Camagni and Capello (2015) referred to as 'territorial capital'. And/or a radical decentralisation of power was required (Martin *et al.*, 2016).

7.6 Conclusion: The UK is Still European

In the pursuit of 'global' opportunities, whether real or otherwise, it needed to be remembered that over two-thirds of the tourism flow in and out of the UK had been with EU countries and was likely to remain so for the foreseeable future. Considerable effort would need to be directed to ensuring that travellers to and from the EU continued to do so as freely as possible (Tourism Alliance, 2017: 14–15).

Yet even without visa requirements, it would not be logistically possible for the UK Border Force to process a further 24m visitors per year through existing non-EU immigration channels. If the UK introduced a reciprocal scheme to the EU's ETIAS requirement for EU nationals, it could impose a significant impact on the UK travel and tourism sectors.

Domestically, although the government's expressed desire for a better distribution of tourism activity across the UK might assist some local cultures, environments and economies, it would be only one of a number of instruments that would be needed to address spatially-expressed social and structural inequalities experienced across the UK, inequalities which in the short term at least, were likely to be exacerbated with the loss of EU funding and support.

And such inequalities would, of course, impact on the production and consumption of tourism-related mobilities at all levels.

Section C

Implications

8 Environmental Implications

This chapter addresses the direct and indirect environmental implications of the UK's exit from the EU for the tourism and hospitality sectors within a wider context of ecological concern.

Invited Contribution

8.1 The Environmental Implications of *Brexit* for Tourism

C. Michael Hall

Brexit represents a major change in the national and international business environment with substantial implications for tourism (Boyle & Wood, 2017; Lim, 2018). Just as significantly, *Brexit* represents a major direct and indirect impact on the environment which will have flow-on effects for tourism given the extent to which the physical environment acts as both a backdrop and a drawcard for domestic and international tourists. Despite the continued uncertainty over the policy settings to be in place following the UK's exit from the EU, the following discusses some of the potential implications for tourism of changes in the environmental conservation and management regimes.

Importantly, and perhaps under-appreciated in discussions on the environmental effects of *Brexit*, impacts from Britain's withdrawal from the European Union will affect not only the countries of the United Kingdom, but also a number of overseas dependencies and territories, those remnants of Empire, such as the Falklands, Gibraltar (see Chapter 12) and Pitcairn, that are spread throughout the world (Amoamo, 2018). Nearly all of these are also strongly focused on the relationship between tourism and the environment as a means of environmental conservation and economic development. For example, in the case of Pitcairn Island, the creation of one of the world's largest marine reserves together with EU funding was regarded as laying a foundation for the island's tourism attractiveness (Clegg, 2017).

Britain's leaving the EU may also have other, unanticipated, affects with respect to international conservation diplomacy and the UK's declining influence on international affairs. In May 2019 the UN Assembly overwhelmingly backed a motion condemning Britain's occupation of the Chagos Islands in the Indian Ocean and demanding the reunification of the Islands with Mauritius (Bowcott an Borger, 2019). In the 116–6 vote against the UK, the only EU country to oppose the motion was Hungary. Although much of the media focus was on the implications for the US military base at Diego Garcia, it should be noted that the Chagos Marine Protected Area declared in 2010 and covering an area the size of France, has been declared illegal by the Permanent Court of Arbitration, and would also be returned to Mauritius, which country wishes to make it open for fishing and tourism (Bowcott & Jones, 2015).

While acknowledging the significance of the above examples, this contribution primarily focuses on the effects of *Brexit* on the UK environment.

Debate on the environment within the mainstream *Brexit* debate tended to focus on the role of EU regulations and directives in supposedly being a constraint on business. However, environmental policy actors tended to portray the UK's relationship with the EU in a positive light with respect to the benefits for the environment. For example, the Green Alliance (2016), a group of environment and conservation experts, including four former chairs of UK environment agencies, argued:

> Britain's membership of the European Union has had a hugely pos-itive effect on the quality of Britain's beaches, our water and rivers, our air and for many of our rarest birds, plants and animals and their habitats. Being part of the Union has enabled us to co-ordinate action and agree policies that have improved our quality of life, including the air we breathe, the seas we fish in, and have protected the wildlife which crosses national boundaries. Higher European manufacturing standards for cars, lights and household appliances have lowered con-sumer energy costs, and stimulated business innovation … We there-fore conclude that *Brexit* would be damaging for Britain's environ-ment. (Green Alliance, 2016)

The environmental implications of *Brexit* for tourism highlights – as one of the underlying themes of this volume – the way in which pol-icies and policy arenas that are ostensibly non-tourism in focus can have an enormous impact on tourism development, attractions and flows. In the case of the environment, concerns such as biosecurity, environmental impact assessment processes, biodiversity and landscape conservation practices (Grant, 2016), can affect tourism substantially, along with the more recognised areas of protected area management and establishment, such as national parks (Frost & Hall, 2009), and protected species conservation, for example the conservation of charismatic fauna and flora that may serve as the basis for ecotourism ventures (Hall *et al.*, 2011).

Core to understanding the place of the environment in *Brexit* negotiations and, consequently, the post-*Brexit* situation, is the significance of the trade agreement between the UK and the EU, as some items of EU environmental law are inextricably tied to the single market, while others are less so. Indeed, the issue of 'alignment' is central not only to the post-*Brexit* situation of Northern Ireland (Chapter 9) but to the environment as well. As the joint EU/UK Agreement of December 2017 (European Commission, 2017a) stated

> The United Kingdom ... recalls its commitment to preserving the integrity of its internal market, and Northern Ireland's place within it, as the United Kingdom leaves the European Union's Internal Market and Customs Union. (Para.45)

> In the absence of agreed solutions, the United Kingdom will maintain full alignment with those rules of the Internal Market and the Customs Union, which now or in the future support North-South cooperation, the all-island economy, and the protection of the 1998 Agreement. (Para.49)

The notion 'full alignment with those rules of the internal market' clearly implies that some EU rules, but not all, will have to be followed, should the UK wish to have full access to the EU. Importantly, such issues of alignment apply not only to the pre-*Brexit* situation but any future changes in EU legislation (House of Commons Library, 2018).

Haigh (2018) argues that EU environmental legislation can be divided into five broad headings when considering the extent of the connection to the single market. As he notes, some items are clearly 'single market measures', others appear to have nothing to do with the single market but can distort competition and so indirectly affect the single market, while some items, such as EU regulation concerning chemicals, fall under more than one heading. These headings are:

A. Standards for traded products
B. Operational standards
C. Procedural standards

D. Quality standards

E. Standards remote from the single market.

There is a long list of traded products subject to EU regulations, and although none of these can be regarded as tourism-specific, the list includes significant items such as motor vehicles. Operational standards include emission standards. These are significant for single market access as they have the potential to distort competition by affecting production costs. Although, in the case of the EU Emissions Trading System (EU-ETS) relating to carbon emissions, Haigh (2018) believes that special legislation will be required.

8.1.1 Procedural standards and impact assessment

Procedural standards are a major area of interest for tourism planning because procedural standards affect the nature of the environmental and social impact assessment of tourism developments, and the characteristics of the assessment and consultation required have the potential to affect competition as a result of the time period required for consultation as well as what is included in any assessment.

Environmental impact assessment (EIA) is significant for tourism in the UK in terms of both assessment of large tourism-related developments, for example airports and transport infrastructure (Hands & Hudson, 2016), as well as developments that may affect tourism, such as wind farms (Phillips, 2015). The legal context for EIA in the UK is already highly complicated, being devolved to the various regions and with legislation for EIA being spread across different sectors (Glasson *et al.*, 2012).

The post-*Brexit* shape of assessment procedures remained unclear, as the nature of EIA, along with social impact assessments and other environmental legislation and regulation, will almost inevitably be tied into the eventual trade deal the UK signs with the EU (Haigh, 2018). For example, one of the conditions of membership for the European Economic Area (EEA) is an obligation to cooperate on the environment, which is the reason why EEA countries have to comply with the EU's EIA directives, implementation of which is overseen by the European Free Trade Association (EFTA) Surveillance Authority (ESA). This situation would be largely unchanged from present EU membership, although some EU directives covering nature conservation fall outside the EEA Agreement (e.g. the Habitats Directive and Birds Directive) (Bond *et al.*, 2016), but may still be significant for single market access discussions.

A separate trade agreement between the UK and the EU was also mooted but the place of the environment in such an agreement is usually unacknowledged. Nevertheless, it remained possible that the EU could insist on stipulations related to EIA in any such agreement. For example, given that a 'level playing field' is a major concern in trade negotiations (Nesbit & Baldock, 2018), it is important to note that the EIA Directive (Council of the European Communities, 1985) was originally adopted based on articles 100 and 235 of the *Treaty of Rome* (Bond *et al.*, 2016). Article 100 provided for secondary legislation, such as EU directives, to be adopted to enable the functioning of the common market. Given concerns from within the UK regarding a deliberate lowering of labour and environmental standards to enable cost competitiveness of UK products and services post-*Brexit* (Green Alliance, 2016; House of Commons Library, 2018), it is not surprising that the EU has sought to ensure that the environment is part of any post-*Brexit* trade agreement.

8.1.2 Quality standards

The two main areas of quality standards relevant to the tourism and the environment relationship are with respect to air and water quality. Ambient air quality can be a factor in the relative attractiveness of destinations (Zhang *et al.*, 2015), and poor air quality is already a significant urban and environmental policy issue in the UK (Carrington, 2019). Although one of the main contributing factors to air quality is road transport, particulates can also be carried between the UK and the EU depending on prevailing weather conditions. In addition, different controls on emissions and other forms of pollution can potentially distort competition. Rivers constitute an important tourism resource and may be attractions in their own right (Steinbach, 1995). Although, with the exception of the Northern Ireland/Republic of Ireland border, the UK is separated from Europe by sea, any pollutants from UK rivers have the potential to affect bordering EU coastal member states. Similarly, as with the case of air quality, any lowering of river quality standards could be construed as affecting competition. It is therefore unlikely that the EU will accept lower standards for water and air quality in the UK (Haigh, 2018).

8.1.3 Standards remote from the single market

EU directives on birds, habitats, protected areas, and bathing water (Blue Flag) are usually not considered as single market measures although are of considerable significance for tourism. Nevertheless, even though the UK may wish to have similar freedoms to members of the EEA with respect to setting whatever standards it likes for nature protection or bathing water quality, it is likely that the EU would resist any attempts by the UK to set lower quality standards. For example,

> ... even nature protection legislation can indirectly affect the single market, as it can prevent a manufacturer building on a protected site. Meeting the EU bathing water standards involves adequate treatment of sewage which entail costs which fall on both the public and industries. Low standards accordingly can provide a competitive advantage. (Haigh, 2018: 12–13)

The UK has argued that it seeks to have greater flexibility in the future in the setting of its environmental legislation and regulation. Environment Secretary Michael Gove argued that the Habitats and Birds Directives impose unnecessary constraints, for example on house building (comments reported in *The Independent*, 25 March 2017, cited in Nesbit & Baldock, 2018). Similarly, DEFRA Minister George Eustice argued that Natura 2000 site designation requirements were too onerous. EU Natura 2000 sites are given a much higher level of protection under EU law than sites receiving the highest level of protection under domestic legislation in the UK. As Nesbit and Baldock (2018: 16) note,

> the designation of sites for some widely dispersed species was contentious, and these might conceivably be reversed, particularly in cases where there is a conflict with development objectives seen as politically important.

The EU's nature legislation therefore involves several transboundary impacts, including the overall goal of creating an ecologically coherent European network of protected sites (House of Commons Library, 2018). This means that a number of environmental measures that, at first glance, appear confined to a specific location in one national jurisdiction are, in fact, subject to far wider spatial governance. This concerns not only protected areas, such as national parks, but also protected migratory birds, the conservation of which has economic and regulatory costs and therefore can be interpreted as affecting competitiveness. As Nesbit and Baldock (2018: 14) observe,

> If a precautionary approach is adopted to the application of the transboundary and single market sensitive criteria, then the list of EU environmental measures where there is no difficulty at all with the UK having the flexibility to take a quite different approach is rather small.

8.1.4 Conclusion

On 11 January 2018 the Conservative Government published its 25-year plan for the environment which detailed how it intended to deliver a 'Green *Brexit*' (House of Commons Library, 2018) (see section 8.2 below). However, in general, current UK environmental governance is relatively weak compared to EU standards, especially given its often fractured and devolved nature, while UK commitments to environmental standards are currently presented as policy commitments rather than being implemented via legislation that would ensure those standards are met. These issues are also particularly pertinent from an environmental perspective given the possibility that the UK's environmental standards could potentially be watered down in any future free trade agreements it negotiates post-*Brexit*, thereby potentially affecting not only its relative competitiveness with the EU but also future expansion of the EU's environmental ambition (Nesbit & Baldock, 2018).

Indeed, Erik Solheim, Executive Director of the UN Environment Programme (UNEP) called on UK environment secretary Gove to honour his promise to deliver a 'Green *Brexit*' (see section 8.2 below), and to ensure that the environment would not suffer from the UK's EU departure, stating,

> Any dilution [of environmental standards] and the UK reputation would be damaged. People in government need to make sure that does not happen. We need to make sure they have those standards or improve them, or meet the ones under the European Union. (Savage, 2018a: 18)

Tourism is diffuse throughout both the EU's and UK's environmental legislative and regulatory regimes. Indeed, it is remarkable just how little the role of tourism is noted in environmental conservation legislation, given how often it is mentioned in policy.

Nevertheless, tourism is a major driver for the conservation of some habitats and species as well as being affected when environmental quality is poor, such as in the case of water quality in rivers and on beaches, and air quality, for example in heritage areas.

Although not so widely recognised in the wider public discourse on *Brexit*, the environment is an extremely important part of the negotiations of the post-*Brexit* relationship between the EU and the UK and is, like much of the *Brexit* discussion, inseparable from the negotiations over the UK's future market access. The future of tourism in the UK, particularly with respect to its environmental governance and the environmental services from which it benefits, is therefore substantially wrapped up with the nature of the market access agreement that the UK is able to achieve.

8.2 The '25-year Plan'

The UK government repeatedly argued that leaving the EU would not lead to lower environmental standards in the UK. Its 151-page 25-year environment plan launched in January 2018 (UK Government, 2018a), as noted above, expressed awareness of the multiple looming crises, and opined that a healthy environment was essential to happiness, wellbeing and a thriving economy. But it also revealed an at best hesitant strategic approach that suggested a weakening of environmental protection in a number of areas compared to existing EU safeguards. The plan was also short on detail, and vague on how goals would be measured and enforced. Attention was, however, focused at its launch on the need to 'eliminate all avoidable plastic waste' within the designated 25-year period, just as the EU was itself introducing new regulations on plastic waste (Box, 2018; Burns *et al.*, 2018; European Commission, 2018d) (Box 8.1).

Indeed, the EU budget commissioner claimed a levy on plastics could be one way for the EU to fill the €13bn hole in its budget left by the UK's exit. The EU wanted 55% of all plastic to be recycled by 2030 and for member states to reduce the use of plastic bags from 90 to 40 per person per year by 2026.

More directly significant for tourism and travel, the European Commission said it would promote easy access to tap water on Europe's streets to reduce bottled water demand, while new port reception facilities would be developed to streamline maritime waste management and reduce the amounts of waste dumped at sea (European Commission, 2018d).

The UK 25-year plan contained no tangible policy for improving recycling rates. With particular significance for the hospitality sector, there was mention neither of a surcharge on disposable coffee cups – only one in 400 of which was recycled – nor of a deposit-return bottle collection scheme (Vaughan, 2018). Such a scheme in Germany achieved a 98.5% recovery rate for plastic bottles (Fearnley-Whittingstall, 2018). The UK government rejected the criticism that the plan put more emphasis on consumers than on producers.

Four months after the 25-year plan was launched, the UK government published details of its intended 'environment watchdog', which would replace the role of the European Court of Justice (ECJ) in environment matters. The proposed statutory body would be supported by a law merely requiring ministers to 'have regard to' core environmental principles. The apparently constrained powers of such a body attracted widespread criticism for leaving the UK potentially far worse off in terms of likely environmental protection (Perkins, 2018).[1]

The source of these fears was that while most current EU environmental law would initially remain in place after *Brexit* through

Box 8.1 Key elements of the 25-year environment plan

Biodiversity	It acknowledged threats from introduced alien species, pests and diseases but merely proposed to 'work with partners to raise awareness' and to 'encourage the development' of better biosecurity. Local planning requirements could be changed, although 'exemptions may be necessary', and any measures should avoid increasing the burden on developers.
Wildlife habitats	To arrest a decline – c17,000ha are lost to new building annually – DEFRA would *investigate* establishing 500,000 extra hectares of wildlife habitat, acknowledging the need for contiguous and continuous areas to enable corridors of animal movement.
Re-engaging children with nature	It recognised the benefits of this, ignoring the fact that government had cut most of its funding for adventure learning and outdoor education. Access and education were dimensions requiring far more attention.
Flooding	It acknowledged floods were a major risk of climate change in the UK and recommended 'putting in place more sustainable drainage systems', overlooking the fact that the government had imposed a seven-year freeze on implementing legal requirements to do just that.
Marine conservation	By July 2019 a third round of marine conservation zones would be designated, although measures to enforce their objectives appeared weak.
Rewilding	The potential to re-introduce 'missing' species would be assessed. A nature recovery network was proposed employing the 'dynamic management of nature'.
Environment watchdog	A 'world-leading' and independent body would hold the government to account. The government's previous watchdog, the Sustainable Development Commission, was closed in 2010.

Sources: Carrington, 2018b; DEFRA, 2018b; Monbiot, 2018; UK Government, 2018a; Vaughan, 2018.

transposition (see Chapter 2), unless specifically prevented by a future EU-UK trade agreement, as implied by Michael Hall above (section 8.1), a future UK government could weaken it. Indeed, Greener UK (2018), a coalition of 13 environmental organisations, expressed 'serious concerns' over the likely (reduced) level of future cooperation between the UK government and the EU, and the impact this could have on such issues as climate change and air quality.

While the environment ministry (DEFRA) might have had good intentions, other elements of government, notably the Treasury, exerted a constraining hand. What was needed was a much stronger green narrative from other parts of government. This was manifest in that governmental responsibility for climate change had been moved from DEFRA to the Department for Business, Energy and Industrial Strategy (BEIS) and therefore would be excluded from the remit of the 'environment watchdog' (see Table 5.1 above).

Ironically, Jacobs (2018) had suggested employing the model of climate change mitigation for effective enforcement of environmental constraints. Under EU law and the UK's 2008 Climate Change Act, environmental limits had been imposed on the whole economy. Every five years government was required to adopt a legally binding 'carbon budget' which was set 15 years ahead, to provide business time to plan, but which should lie on a trajectory towards the long-term goal of an 80% reduction in UK emissions by 2050, relative to 1990 levels. This, Jacobs argued, should become the model for a new sustainable economy act (also Jacobs & Mazzucato, 2016). Yet the existing UK Committee on Climate Change (CCC), which had only advisory and reporting powers, pointed out that already it was clear the UK would fall short of its 2025 and 2030 carbon reduction targets (UKCCC, 2018).

Subsequently, a 'new' UK air pollution strategy was announced that was both thin on detail and lacking any mechanism for accountability (Harvey, 2019). An amendment to the Climate Change Act committed the UK to reach zero carbon emissions by 2050. Norway had already committed to achieving this objective by 2030, and Finland by 2035.

UK public opinion appeared to support the maintenance of EU standards. In a survey of 2,004 people undertaken in January 2018 by *Opinium* for the Institute for Public Policy Research (IPPR, 2018), 61% wanted to retain or tighten standards on vehicle fuel emissions, 74% expressed the same opinion for renewable energy targets and 85% for consumer cancellation rights (Table 8.1).

As part of responses to a wider range of EU standards and likely post-*Brexit* UK policy, the IPPR (2018) argued that the UK public expressed little desire to deregulate, in response to those politicians arguing for deregulation on the grounds that the UK would not have the resources to keep up with EU regulations after exiting. Strong support

Table 8.1 Selected summary results of IPPR/*Opinium* survey on EU standards (% rounded)

Rules	Vehicle fuel emissions	Renewable energy targets	Consumer cancellation rights
Stay the same	28	26	58
Stricter/tightened	39	48	27
Relaxed/loosened	7	4	4
Removed altogether	4	4	2
Don't know	22	17	10
Total: 2,004	100	100	100

Source: IPPR, 2018.

for maintaining EU-derived standards was evident in both *Remain* and *Leave* voters.

Within two weeks of the 25-year plan being launched, the EU's environment commissioner gave an ultimatum to the UK and eight other member states to show how they would comply with EU air pollution laws or face legal action at the ECJ (Neslen, 2018). Levels of nitrogen dioxide, mostly produced by diesel vehicles, had been above EU legal limits since 2010 in the majority of Britain's urban areas, and this was estimated to cause 23,500 premature deaths every year (DEFRA, 2016). In September 2017 the UN's special rapporteur on pollution said the UK government was 'flouting' its duty to protect its citizens. Along with five other EU member states the UK was indeed referred to the ECJ having failed to deliver 'credible, effective and timely measures to reduce pollution as soon as possible, as required under EU law' (European Commission statement quoted in Carrington, 2018a: 8). The problem had been recognised and declared a public health emergency by a cross-party committee of MPs in 2016. This did not bode well.

If the UK government could not protect its own citizens, what did this signify to the UK's (potential and actual) visitors?

8.3 Rural Environment and Tourism

While the significant growth of UK rural tourism and recreation helped to focus attention on the well-being of Britain's rural and semi-rural environments, debate on post-*Brexit* rural issues inevitably adopted a wider frame.

Although praising much of the work undertaken by the EU in raising environmental standards in the UK, Monbiot (2016b) cited two 'astonishing idiocies' perpetrated by the Union: the emphasis placed upon replacing existing transport fuel with biodiesel – itself generating substantial greenhouse gas emissions and contributing to the destruction of Asian rain forests for the growing of palm oil – and the

CAP, which, providing £3bn annual subsidies for UK farmers, was seen to act as a 'perverse incentive for clearing wildlife habitats' (Monbiot, 2017: 31).

Farm subsidies made up 38% of the EU budget, with 80% of the subsidies going to just 20% of Europe's farmers, or rather, farm-industrial complexes. These 'basic payments' related to land area, reinforcing the dominance of large farms and 'efficient' production, where attractive landscapes had often been reduced to the large, hedge-less fields of mono-culture. Further, in the UK, farmers could not receive basic payments for land featuring ponds, wide hedges, salt marsh or regenerating woodland. The EU's independent auditor, the European Court of Auditors, concluded in December 2017 that existing 'greening' payments were unlikely to provide significant benefits for the environment and climate because they changed practices on just 5% of EU farmland (Barkham, 2018).

Brexit presented the perceived opportunity to create a new agri-cultural policy that could better help restore natural environments. Monbiot (2016b) among others, argued that CAP basic payments should be replaced by a rural hardship fund and an environmental protection fund that more sensitively paid for wildlife and habitat restoration and flood prevention schemes. Indeed, at the farming conference where an independent Nature Friendly Farming Network was launched (National Trust, 2018; www.nffn.org.uk,) environment secretary Gove insisted that he would dismantle the £3bn a year farm subsidy scheme once the UK was removed from the EU's Common Agricultural Policy, replacing rewards for simply owning land with support for enhancing nature in order to invest in a series of 'public goods':

> … planting woodland, providing new habitats for wildlife, increasing biodiversity, contributing to improved water quality and returning culti-vated land to wildflower meadows or other more natural states. (Preston, 2018: 35)

Given the importance that landscape can play not just in terms of visitor (passive) appreciation and (active) consumption but as an element of national identity (see Chapter 4), Gove's approach was viewed as representing a belief that 'the countryside' could provide common ground upon which the factions of a post-*Brexit* divided UK might meet (Preston, 2018).

While the UK government was committed to sustaining basic payment subsidies until 2022, pressure groups such as the National Trust and the Campaign to Protect Rural England, argued that this must be accompanied by:

- clear guarantees for farmers that food and environmental standards would be maintained or strengthened;

- ensuring £800m of greening subsidies were redirected in 2019 into more effective incentives rewarding farmers for working in unison with the natural environment (Ghosh, 2017); and
- assistance for the farming sector to become more resilient and sustainable, not least through encouraging farm size diversity to support rural communities particularly in marginal areas (Willis, 2017).

In the shorter term, however, *Brexit* presented two difficult issues for agriculture. First, there were 86 agricultural products likely to be subject to tariff rate quotas, each of which the UK, as 'Global Britain', might need to negotiate with other trading nations. Second, as noted in Chapters 5 and 6, farm and rural hospitality workers from other EU countries were leaving as a result of the referendum outcome and the depreciation of sterling, resulting in critical labour shortages (Monbiot, 2017).

8.4 Conclusion

Environmental law in the UK changed considerably during the four and a half decades of EEC/EC/EU membership, and it was hoped that most of the innovations introduced through those institutions would be retained. There was, however, likely to be a wish to restore more discretion over the outcomes to be achieved as opposed to having strict obligations to satisfy targets and standards.

Most environmental matters would be the responsibility of the devolved administrations. The UK's freedom of action would continue to be constrained by obligations in international law, including those establishing a new relationship with the EU.

In structural terms, the most significant changes were likely to be the loss both of the stability provided by the slow processes of making and changing EU law and of the means to call to account the UK and devolved governments' performances in meeting their environmental commitments (Reid, 2016: 407).

The UK's withdrawal from the EU posed significant direct and indirect environmental implications for the tourism and hospitality sectors, not least in the extent to which the physical environment acted as a stage setting, a symbol of national imagery and an intrinsic attraction and health benefit in its own right for domestic and international visitors and recreationalists.

Meanwhile, the EU's draft 'strategic agenda' for 2019–24 (Government Europa, 2019) was criticised as lacking ambition in tackling climate change and species extinction (Rankin, 2019). Notable environmental campaigner Greta Thunberg (2019) told the EU it needed to 'double' the ambition of its climate targets after the European

Commission (2018m) had called for the EU to be 'climate neutral' by 2050. The 'Thunberg Effect' was raising public awareness of pressing environmental threats and helping to expose the 'weaponisation of ignorance' (Mann, 2019; Mann & Toles, 2016) embraced by climate change deniers.

Note

(1) Between 2003 and 2016 the European Commission began 753 actions against the UK government, of which around 120 related to environmental matters; only 29 of the latter actually reached the ECJ (Hogarth & Lloyd, 2017).

9 Inconvenient Cross-border Mobilities I: Ireland

To much of Europe, Brexit appears to be an exercise in British self-harm, which it is. But in Ireland Brexit is potentially lethal too ... [a hard border] ... would be a gratuitous act of hostility towards the Irish economy and people ... that would be unforgivable.

Anon, 2017b: 26.

9.1 Context

The Republic of Ireland's relationship with the UK is unique among EU member states. Its intense historical and cultural links differ from, and are arguably stronger than those between any other EU member state and the UK. The shared, often contested history of colonialism, migration and division contextualises the relationship like no other within the EU. As such, tourism and related mobilities to and within the island of Ireland faced considerable potential impact from the UK's EU withdrawal (e.g. Bergin *et al.*, 2016).

Crucially, the Republic of Ireland is the only EU member sharing a land border with the UK (499km/312ml long) (Figure 9.1). The border is replete with psychological and emotional meaning. Its physical presence has cut through homes and businesses, separated families and interceded between producers and their markets. In recent years the border's invisibility has been essential for the smooth running of economies and communities on and between both sides. Indeed, the 1998 Belfast/Good Friday Peace Agreement that established power sharing in Northern Ireland (see Box 9.4, p. 144) has depended on the continued existence of an invisible border (Boyle & Wood, 2017).

The UK government's determination to leave the EU's single market and customs union threatened to reconstruct the UK-Ireland border as a 'hard' EU external border. And while Northern Ireland's Democratic Unionist Party (DUP) could view this as an opportunity to lessen the influence of the Republic and republicans in the North,

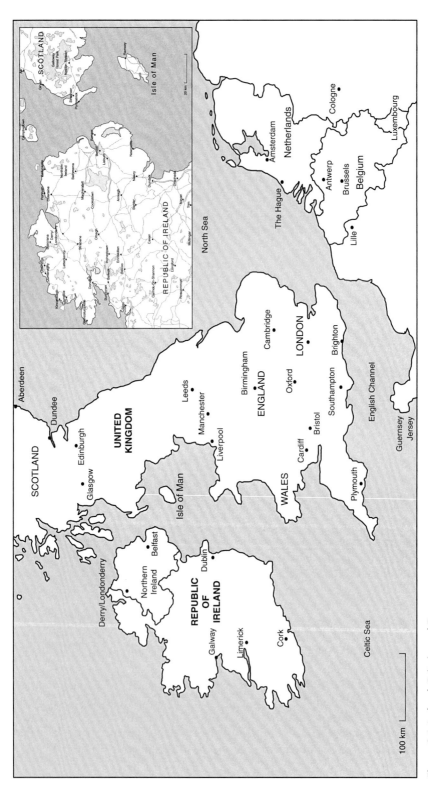

Figure 9.1 Ireland, Britain and Europe

… re-establishing a geopolitical border symbolic of a nationalist strug-
gle that has taken 800 years to finally [sic] get to a relatively peaceful,
stable and European future. (Wilson, 2016: 498–499)

would not only be an Irish tragedy.

With an estimated 110m person-crossings annually between the
Republic and Northern Ireland (DftE, 2018) (Box 9.1), the border is
currently wholly open to the significant tourism and recreation, shopping,
migration, commuting and trade flows continuously crossing it.

The border's political significance has tended to deflect attention
away from the nature of the border itself (Box 9.2) and the ways it was
constituted and experienced. Both elements became central in the role
the island of Ireland played in the unravelling of the UK's exit strategy
from the European Union.

As a forewarning, in their study of the border and its customs
arrangements between 1921 and 1945, Nash *et al.* (2010) drew attention

Box 9.1 Cross-border issues I: numbers crossing

There is no one complete and definitive data source that records the
number and purpose of all cross-border movements.

Traffic count data from 15 Northern Ireland and Republic border crossings provides an average figure of 118,000 daily vehicle crossings.
Since there are many more crossing points than those with traffic
counters, the actual number of border crossings is likely to be substantially higher. The traffic count figures equate to approximately
43m annual cross-border vehicle trips. The traffic flow in May 2017
comprised 87% cars, 7% light goods vehicles, 5% heavy goods vehicles and 1% buses and coaches.

The Northern Ireland transport model covering all cross-border movements has estimated that there are 143,000 car and light good vehicles crossings each (busiest) 12-hour period and an average of 19,600
HGV crossings.

The vehicle crossings equate to 243,000 person-crossings over the
12-hour period each day, when standard vehicle occupancy rates are
applied. This gives an annual estimate of 89m person-crossings for
the 12-hour period each year. Extrapolating this figure for a 24-hour
period provides an estimated annual total of 110m person-crossings.

Translink, the main provider of public transport in Northern Ireland,
estimated that in 2016/17 there were approximately 900,000
cross-border coach passenger journeys and 868,500 cross-border rail
passenger trips.

Northern Ireland residents undertook more than 1.2m journeys departing
from Dublin airport in 2015.

Sources: ABTA, 2017; DftE, 2018.

Box 9.2 Cross-border issues II: crossing points

According to the first officially agreed count since the island was parti-
tioned in 1922, there are 208 border crossings.

A joint mapping exercise involving the Republic's Department of
Transport and the Northern Ireland Department for Infrastructure,
begun in 2017, faced a number of difficulties in trying to definitively
map all roads, paths and dirt tracks that traverse the 499km/312 mile
dividing line that itself was based on local government area boundaries as they existed in the 1920s.

The border runs along the middle of eleven roads, which is more than
twice the number originally believed, it passes through the middle of
at least three bridges and dissects two ferry crossings.

One problem the joint exercise faced was over different definitions
employed by Belfast and Dublin in classifying whether roads were
public or private.

Source: Hutton, 2018.

to the ways in which the state was continuously remade and experienced
through practices, texts and techniques of regulating the movement of
people, animals and objects. The authors' historical analysis reflected
how the political entered everyday life, not just through the ideological
dimensions of partition, but especially through the ways in which
'borderlanders' were required to come to terms on a quotidian basis with
the existence of the customs barrier and to suffer its inconveniences,
inconsistencies and illogicalities.

9.2 Ireland: Tourism Dimensions

Prior to the 2016 referendum vote, the UK was the Republic of
Ireland's largest inbound tourist market, representing around 40%
of total visitors to the Republic and more than 20% of 'overseas'
tourism income (Edwards, 2018). UK tourism comprises both business
and leisure travel with a significant if largely undocumented VFR
component. UK visitor numbers peaked in 2006 at almost 4.1m. Impacts
of the subsequent economic downturn saw this figure dip to just under
2.8m in 2012, only to recover to 3.9m by 2016. However, following the
2016 UK EU referendum result, sterling's effective devaluation rendered
the Republic a more expensive destination.

UK visitors declined to 3.2m in 2018 (ONS, 2019b: Table 3.10). The
expectation within the sector was that the UK market would continue
to decline, although at what pace and level was difficult to predict,
depending upon whether border controls would be re-introduced and the
extent of other *Brexit* impacts such as a slowing UK economy.

This reduction in the number of UK visitors to the Republic has been compensated for by significant increases in other markets. Over the period 2007–2017, while the UK visitor market to the Republic declined by 8%, North American visitor numbers increased by 96%, those from 'other Europe' by 34%, and 'rest of the world' visitors increased by 95%. In 2017, 'other Europe' accounted for 3.5m visitors and North America for 2.1m. It was appreciated, however, that EU markets would be impacted by *Brexit* as those countries would be encouraging their own domestic tourism to fill any gaps left by fewer UK visitors.

The implications for Ireland would therefore not just stem from the UK, but from the ripple effects of *Brexit* (Murphy, 2016a), not least for those countries whose visitors had hitherto gained access to the Republic via the UK. Access to the UK via Ireland, however, offered an intriguing area of possibilities.

Dublin Airport recorded a 1% increase in passenger numbers to and from Britain in 2017 to almost 10m. While weaker sterling contributed to a decline in traffic originating from Britain, increases in both Irish outbound and in transfer traffic to UK destinations compensated for this (daa plc, 2018: 23).

Further, more stringent passport control between the UK and other EU countries could be positive for Ireland as the Republic enhanced its position as a gateway for long-haul connecting flights. Hitherto, the UK had been an important access point to Ireland for many long-haul destinations such as China. While *Brexit* could hamper this flow of visitors from such countries, it offered the opportunity for establishing more direct long-haul routes to Ireland.

Long-haul passenger numbers to and from the Republic increased by 19% in 2017 to 4.3m, 3.5m of those being trans-Atlantic. Passengers using Dublin airport as a hub rose by 32% to 1.8m (daa plc, 2018: 23). For 2018, *Aer Lingus* added new summer services to Dublin from Seattle and Philadelphia, while Air Canada offered new access from Montreal to Dublin and from Toronto to Shannon.

In revenue terms, losses to the Republic's economy from a decline in UK visitors had been largely offset by gains from these other regions, which delivered a higher spend per capita and a longer average stay. In 2017 (Q1-3) UK visitors had an average spend of €273 compared to €517 for other European visitors, €730 for North Americans and €931 for visitors from other areas. Indeed, the North American market overtook the UK market in terms of generated income, increasing total spend from €1bn in 2014 to €1.7bn in 2017, while the UK market only grew from €1.3bn to €1.5bn over the same period (Murphy, 2018).

Significant growth in the Asian market was anticipated following the June 2018 introduction of direct flights to Dublin from Hong Kong (Cathay Pacific) and Beijing (Hainan Airlines) (Tourism Ireland, 2018b: 9). *Fáilte* Ireland (the Republic's tourist board) had been working

Table 9.1 *Tourism Ireland:* a slightly schizophrenic symbol

Tourism Ireland: Role and context

Function	To promote overseas the whole island of Ireland as a tourist destination.
Administration	Tourism Ireland Limited is a publicly owned limited company operating under the auspices of the North South Ministerial Council (NSMC) through the Department for the Economy in Northern Ireland and the Department of Transport, Tourism and Sport in the Republic of Ireland.
Genesis	The North South Ministerial Council (*An Chomhairle Aireachta Thuaidh Theas*) was established under the Belfast/Good Friday Agreement (1998), to develop consultation, cooperation and action within the island of Ireland concerning agriculture, education, environment, health, tourism, and transport.
Role of the NSMC in tourism	The Council meets to take decisions on policies and actions to be implemented by *Tourism Ireland*.
Reality	The organisation operates across two countries with different currencies, statistical sets and political priorities. While *Tourism Ireland*'s remit is the island of Ireland as a whole, its language can sometimes be confusing and possibly misleading: e.g. 'The fall in the value of sterling has made holidays and short breaks *here* [italics added] more expensive for British visitors' (Tourism Ireland, 2018c). See also Table 9.2.
Online presence	International website is www.ireland.com, with 29 market sites available in 11 language versions attracting 19.3m page visitors in 2017.

Sources: North South Ministerial Council, 2018; Tourism Ireland, 2018a, 2018c.

with hotels and tourism businesses on a 'China Welcome Programme' in advance of this significant (bypassing the UK) development. But marketing efforts needed to continue focusing on the wider global arena in order to sustain post-*Brexit* market diversification.

Overseas marketing is undertaken by *Tourism Ireland* which acts to promote the whole island of Ireland, a somewhat delicate political task (Table 9.1).

The Northern Ireland market is an important leisure segment for the Republic's border counties, with hotels in these areas reporting strong business from the North up to 2016 while sterling was relatively strong. In a survey undertaken three months after the UK EU referendum vote (Murphy, 2016b), 82% of Republic border county hoteliers reported that their business was feeling the impact of the vote and subsequent sterling devaluation; 75% expected an overall decline in demand levels over the next 12–15 months. This was largely borne out.

However, after the UK's withdrawal, Ireland would be the only native English-speaking country within the EU, which could have a positive effect on the conference market, with (compared to the UK) easier travel and a unified currency within the eurozone. The 2016 International Conference and Convention Association rankings (ICCA, 2017: 20) placed Dublin 13th among world cities for hosting international association conferences, with 118 meetings that year.[1] The UK's becoming a more difficult or expensive conference destination would benefit Dublin and the Republic in this lucrative sector.

Table 9.2 *Tourism Ireland:* 2018 trends

Visitor markets to the island of Ireland	January–March 2018 visitor numbers growth
Overall overseas	+7%: 124,500
Mainland Europe	+14% Germany +30.6% Italy +18.3%
North America	+13%: 38,300
Impact of tourism in Ireland	More than 4% of GNP "in the Republic of Ireland" employing 281,000 people across the island
Tourism in 2016	10.3m overseas visitors to the island of Ireland (9% increase over 2015), delivering revenue of €5.3bn (also a 9% increase). ('Mainland') Britain 47%, USA 13%, Germany 6% and France 5% represented almost three-quarters of overseas visitors
Tourism in 2017	10.65m overseas visitors: revenue €5.78bn.
Tourism in 2018	*Tourism Ireland* aimed to increase overseas tourism revenue by 5% to €6bn for the island, with at least 10.8m visitors.

Sources: Tourism Ireland, 2017, 2018b, 2018c.

In 2017, Crowe Horwath partnered *Fáilte* Ireland to develop an online 'Brexit Readiness Check', a self-assessment tool, to help Republic tourism businesses assess their level of readiness for the UK's EU withdrawal (www.getbrexitready.com). Launched in September 2017, by March 2018 a number of trends had been revealed through survey (sample size not given) (Table 9.3).

The threat of *Brexit* had been viewed as a catalyst for tourism businesses to review and update their sales and marketing strategy, to diversify and target new markets. Although it would be expected that indicators of business readiness would improve in the short- to medium term, Table 9.3 suggests that the requirement for future planning and preparation was

Table 9.3 *'Brexit* readiness' among Irish tourism businesses

Indicator of *'Brexit* readiness'	% of sample (apparently rounded)
Businesses having a plan to deal with *Brexit*	50
Businesses that had carried out a *Brexit* impact assessment	<10
Businesses that monitored and reviewed business performance in relation to a sales and marketing plan	70
Businesses welcoming greater collaboration with other businesses in their area	90
Businesses lacking confidence in understanding the visitor needs and preferences of non-UK markets	33
Businesses having an entry strategy for new markets	50
Businesses that would either maintain or increase their marketing spend in the UK market	64
Businesses having improved some of the areas where they could reduce costs and improve competitiveness	50

Source: Murphy, 2018.

considerable (Murphy, 2018). *Fáilte* Ireland was providing training and support programmes to assist businesses in preparing their strategic and operational responses to the threats posed by *Brexit*.

While having an impact on UK travel to the Republic, particularly in the border counties, sterling's 2016 devaluation further stimulated cross-border shopping trips from the Republic to the North (already significant since at least the 2007+ financial crisis), to some extent draining Republic border economies while exacerbating congestion and potential local inflation in such near-border honeypots in the North as Newry and Derry/Londonderry. Newry, for example, with a resident population of just 30,000, has seen the development of two large shopping complexes – Buttercrane (with 1,000 parking spaces) and The Quays (1,300 spaces) – on the southern edge of the city centre, closest to the border. On the city's free visitor map (Newry, Mourne and Down District Council, 2017) 15 further designated parking areas within the city are highlighted.

The Irish government emphasised the importance of maintaining a 'soft' border, not least because of concerns over potential job losses due to export restrictions arising from *Brexit*, and their economic knock-on effects. Tourism businesses largely remained positive, with hoteliers projecting a continued increase in international visitors during 2018.

9.3 Related Mobilities

Migratory flows between Ireland and the UK have been a notable feature of the demography and economics of the two islands for centuries, although the proportion of emigrants from Ireland travelling to the UK has decreased over time. In 1987, over a half of Ireland's emigrants were bound for the UK, but this figure had fallen to 22% by 2016, largely for two reasons. First, before the 21st century EU enlargements, almost all emigrants from Ireland were Irish nationals. But following inflows of non-nationals into Ireland, by 2016 almost 60% of 'emigrants' from Ireland were not Irish nationals but non-nationals mostly returning to their countries of birth. Second, some emigration from Ireland due to the recession was to countries which were somewhat insulated from its worst effects, such as Australia.

In spite of the relative decline of the UK as an emigration destination, flows to the UK doubled during the recession. With about 2 million people employed in Ireland, there was an outflow to the UK of 98,000 between 2010 and 2015. These figures suggest that the UK labour market helped to absorb some of the labour which would otherwise have remained unemployed in Ireland. There was also a steady stream of emigration from Ireland to the UK during the years of Ireland's economic boom, although at a lower rate. This suggested that emigration may have been generated for reasons of professional development, facilitated through shared legal, linguistic and other institutional features. Such

Box 9.3 Cross-border mobility potential

375,900 Irish-born people living in the UK
277,200 UK-born living in the Republic
57,000 residents of the Republic born in Northern Ireland
38,000 Northern Ireland residents born in the Republic: 23,000 of those
    live in the 5 district council areas closest to the border (61% of their
    populations)

Source: Travis, 2017.

human capital accumulated in the UK could be recognised and rewarded in Ireland subsequently (Barrett & Goggin, 2010).

Almost 376,000 people who were born in the Republic of Ireland were resident in the UK (according to the 2011 UK census). More recent figures suggest that 277,000 UK-born people are resident in the Republic of Ireland (Box 9.3). As a consequence, high levels of short-term movement between the two islands have been sustained such that in 2016 2.7m passengers travelled between the Republic and British 'mainland' UK) sea ports (DftE, 2018); 41% (3.5m) of arriving air trips and 50% of flights out by Irish residents (3.5m) also involved Britain (ORC, 2016).

Such high levels of movement have been prompted by a range of push and pull factors over the decades, and since 1952 they have been facilitated by the 'common travel area' (CTA: this has also included the Isle of Man and the Channel Islands, which have never been part of the EU). Under this arrangement, Irish and British citizens have been allowed to move between the islands and to settle for work purposes. A person who has been examined for the purpose of immigration control at the point at which they entered the area does not normally require leave to enter any other part of it. Maintenance of the CTA has been a core principle for all Irish governments. Partly in recognition of the CTA, Ireland and the UK were able to jointly opt out of the EU's Schengen arrangement in 1995 (Barrett & Morgenroth, 2016).

Most of the important rights currently available to UK and Irish nationals under the CTA are duplicated in EU free movement/citizenship law. With *Brexit* the CTA (assuming it continues) would exist between one EU member state and one non-member. To date, there have been no examples of an EU member state offering non-EU nationals better status than EU nationals might have. European Union agreement on any continuation of the CTA between the UK and the Republic of Ireland was therefore required in the UK's exit negotiations and satisfactory resolution of a subsequent EU-UK trade agreement (DftE, 2018: 9).

Maintenance of free movement between the islands was identified by the Irish government as a core demand for conditions covering any UK withdrawal from the EU. Dublin pointed to the critical importance of this for the Northern Ireland peace process and for the continued

Box 9.4 Significance of *Brexit* for the Belfast/Good Friday Agreement, 1998

1. The agreement clearly envisaged that Northern Ireland's future constitutional arrangements would be worked out in the context of continuing partnership between the North and the Republic. To remove Northern Ireland from Europe without its consent is a rejection of the fundamental bilateralism of the peace process.
2. The all-Ireland dimension of the agreement was fundamental to securing the support of nationalists. The Republic was able to revise its constitution – to no longer lay claim to the North – for the first time fully recognising the legitimacy of partition. That the island of Ireland can be both one unit (e.g. for overseas tourism marketing (Table 9.2), rugby union and cricket 'national' teams), and yet two separate political entities, has enabled relative peace and multi-dimensional cooperation to be sustained. Most particularly, the creation of a hard border would be contrary to the full recognition of nationalist aspirations enshrined in the settlement.

Source: McBride, 2016.

normalisation of relationships within Northern Ireland, on the island of Ireland, and between Ireland and the UK. Since the signing of the Belfast/Good Friday Agreement in 1998 (Box 9.4) – widely viewed as a 'European' document rather than a British or Irish one (Wilson, 2016) – the border between north and south had been essentially non-existent for most practical purposes (Figure 9.2). This was in stark contrast to the

Figure 9.2 The busy, free-flowing border bridge between Lifford (Republic) and Strabane, April 2018

situation in the 1970s, 1980s and into the 1990s, when, because of the 'Troubles' and despite the CTA, there was a significant security presence along the border and reduced numbers of crossing points.

By 2017, cross-border trade had more than doubled since the Belfast/ Good Friday Agreement, which was based on the key principle that there would be no change to Northern Ireland's constitutional status without majority consent.[2] The majority in Northern Ireland in the 2016 EU referendum voted to 'Remain', by 56% to 44%.

9.4 Tourism-related Trade

The Republic kept strong economic connections with the UK after independence, and when both the UK and Ireland joined the EEC in 1973, 55% of Irish merchandise exports were destined for the UK and 50% of imports came from the UK. Since then, the Republic has significantly diversified its trade and the share of both services and merchandise trade with the UK has been steadily declining. However, the UK still accounted for around 17.2% of total trade (merchandise and services imports and exports), making it the single most important individual trading partner for Ireland, just ahead of the United States (17.1%), while the rest of the EU represented 35.8% (Barrett & Morgenroth, 2016).

While aggregate trade figures indicated the overall exposure of Ireland to an introduction of trade barriers, they masked a significant variability in terms of type of firms, sectors, products and regional exposure, identifying key sensitivities for the country. The agri-food sector was particularly exposed, with 39% (€4.8bn) of its exports to the UK and 47% (€3.7bn) of products sourced from the UK (Boyle & Wood, 2017).

A significant share of Ireland's global exports has derived from the presence of foreign multinational corporations (MNCs) in the country. Indigenous firms have been far more dependent on the UK market, with 43.5% of their exports being destined for the UK, while for MNCs the share has been just 10.6% (Barrett *et al.*, 2015). The vast majority of MNCs are exporting firms and operate in advanced sectors, while a significant share of indigenous firms have been active in more traditional sectors, often tourism- and hospitality-related, such as food and beverages, which have tended to be more dependent on the UK market. Unsurprisingly, trade intensity with the UK has been greatest along the Republic's (economically lagging) border counties.

Almost 20% of Irish services exports are destined for the UK but, again, there are significant differences in the level of dependence across sub-sectors. For example, almost 80% of Irish transport services exports go to the UK, notably in the provision of aviation services; for Irish financial services the UK (32.8%) is a more important destination than the rest of the EU combined (32.1%), highlighting the close integration of the International Financial Services Centre in Dublin with the City of London (Barrett & Morgenroth, 2016).

UK retailers have a significant presence in the Irish market. Many products' wholesalers based in the UK serve both the UK and Irish market, as the Irish market is too small on its own. Trade barriers would result in higher prices and would thus have an inflationary impact on the Republic's tourism and hospitality sectors.

9.5 Non-Policy

The UK government's *Brexit* strategy as it related to Ireland, appeared both paradoxical and poorly thought through. If the UK left the customs union and single market there needed to be an EU external border, potentially constraining trade, prosperity and the movement of people, as well as compromising the Belfast/Good Friday Agreement. But the UK prime minister repeated that the government wished to retain the Common Travel Area. Not only did this provoke an outcry from some 'Leavers' against what they saw as offering a backdoor to immigration, but the position was clearly contradictory (Box 9.5).

The EU's efforts had been focused on putting pressure on the UK to provide a means of achieving the 'seamless and frictionless border' that

Box 9.5 The unresolvable paradox?

1. The UK prime minister, declaring at the outset of withdrawal negotiations that Britain would leave both the single market and the customs union, effectively ruled out a range of possible compromises.
2. Yet in June 2017 she told parliament that post-*Brexit* arrangements for EU citizens would not apply to Irish citizens, and that their rights would continue to be guaranteed under the Common Travel Area arrangement (Quinn & Doward, 2017).
3. When, in November 2017, the UK foreign secretary was asked to reaffirm his February 2016 pledge that *Brexit* would leave Irish border arrangements 'absolutely unchanged', he responded 'there can be no hard border; that would be unthinkable. It would be economic and political madness' (Settle, 2017: 6).
4. Citing paragraph 54 of the Joint Report from the negotiations of the EU and UK, the UK government pointed out that the European Commission had acknowledged that the CTA arrangements could continue. The Irish government was clear in its commitment to the CTA, and the UK government was firmly committed to maintaining the CTA arrangements after the UK left the EU, 'an objective shared by the Crown Dependencies' (UK Government, 2018e: 3).

the UK and Irish government leaders had promised after the referendum vote. Despite protesting that the border issue could not be resolved before establishing a borders policy, a customs agreement and a trade agreement, the initial UK government view was that a 'light touch' 'smart' border could be sustained relying on electronic assistance to have spot checks of vehicles, to use number plate recognition technology, trusted trader status for border regulars, radio frequency identification (RFID) and CCTV cameras to try to monitor the movement of goods across the border. While there are certainly a number of roles that fast-evolving technology can achieve to speed up border crossings – an approach encouraged by an upbeat government-commissioned review on industrial digitalisation (White, 2018) – some critical cross-border goods, such as animals and animal feed, highlighting issues of biosecurity, would need to be physically checked under EU law, requiring a customs presence.

Border communities that had flourished since the 1998 Belfast/ Good Friday Agreement campaigned for the open border to continue after *Brexit* (Box 9.6, Figure 9.3). The only way to achieve this was for

Box 9.6 Border Communities Against Brexit (BCAB)

An organisation seeking to protect the people who lived and worked close to the border in both the Republic and Northern Ireland including representatives of vulnerable adults, SMEs, farmers, hoteliers and others in the tourism sector, out of a concern that the North's majority 'Remain' vote would not be respected.

Days of action included lobbying government in London, Dublin, Belfast and Brussels, and setting up protests at mock border post checkpoints.

In July 2017 the group was awarded a European Citizen's Prize.

– 'A hard border in Ireland would create real hardship for people in this region who cross the border on a daily basis. We could very well be facing customs checkpoints, traffic delays and the closure of local border roads.'

– '...the end of EU economic support for peace-building projects and underfunded border areas will hit many vital community projects.'

– 'New restrictions on cross-border agricultural trade would be a devastating development for farmers, particularly in border counties. Farming communities, north and south, have always worked together.'

Alongside the group's slogan 'No EU frontier in Ireland' it employed distinctive imagery on its posters and banners, including pictures of long queues of lorries and the strong emotional symbol of a silhouetted watchtower (Figure 9.3).

Sources: Anon, 2016; Edwards, 2016; European Parliament Liaison Office in Ireland, 2017; RTE, 2016, 2017.

Figure 9.3 A dramatic 'Border Communities Against Brexit' poster near Strabane

Northern Ireland to remain in the European customs union, but this notion was rejected by the province's Democratic Unionist Party (DUP), upon whose votes the UK Conservative government depended following the June 2017 general election (Chapter 2). As unionists, the DUP insisted on 'the single market of the UK'.

The Republic was ranked second in the EU for ease of customs procedures. In 2016 approximately 1 million roll-on roll-off units arrived, over 90% from the UK, with around 1 million HGVs and 1.3m LGVs moving in each direction across the land border (Cody, 2017). The interdependence of supply chain activity either side of the border is symbolised by the iconic Irish beer/stout Guinness, which travels from Dublin to Belfast for bottling and back to Dublin for export to Britain.

Customs checks would inevitably exert a negative impact on trade flows and delay the release of goods, with administrative and fiscal burdens on traders who had no experience of such constraints. In 2017 there were 12,000 businesses exporting to the UK from the Republic and more than 60,000 importers (Cody, 2017).

A confidential study undertaken by the UK government and the European Commission *Brexit* task force identified 142 cross-border activities that would be negatively affected by a 'hard' *Brexit* (O'Carroll, 2017b). These ranged from agreements on mobile-phone roaming charges that limited charges to local rates across the whole island, improved social cohesion in sports and cultural activities, to joint health services including heart surgery and cancer treatment: emphasising the depth of cooperation that had developed since 1998.

While cross-border traffic is significant, in terms of trade, greater activity for both Northern Ireland and the Republic is conducted across the Irish Sea. Significant *Brexit* challenges would be faced at Dublin port, which handles most of the goods from the North and the Republic exported to Britain and freight to and from the rest of the EU. With large numbers of passengers passing through both Dublin and Rosslare ports to and from mainland UK, existing physical infrastructure and traffic streaming appeared inadequate for any increased demand to support an EU external border placed upon customs control capacity (ORC, 2016).[3] In response, by the time of the UK's first 2019 exit date, the port had spent €30m and set aside 7.8ha of land for such purposes. 'Hundreds' of newly recruited or assigned revenue officers, police, customs inspectors and vets were said to be on standby (Carroll, 2019).

9.6 Resolution?

Within the March 2018 interim draft agreement on withdrawal, over which, in fact, there had been only '75%' agreement between the UK and EU, the question of the Irish-UK border remained one of the unresolved areas. In the event of no agreement being reached on exit negotiations, until a UK-EU trade agreement was resolved, Northern Ireland could remain in regulatory alignment with the EU. In other words, the EU-UK border would run through the Irish Sea and not across the island of Ireland.

Although this 'backstop' position[4] was lightly considered by some as merely involving a few Irish Sea ports, in practice it would embrace ports in the Irish Republic, Northern Ireland, the Isle of Man, Scotland, Wales and England. Upholding the inviolable and indivisable 'single market of the UK' was a dismissive DUP response that complemented the UK government's initial rejection of this EU proposal (Johnson *et al.*, 2018), and parliament's successive rejections of it in 2019.

Nonetheless, an *Ipsos MORI*/Queens University Belfast sample survey of 1,015 Northern Ireland residents conducted in September 2017 found that 59% of Unionist voters and 64% of 'Leave' voters agreed that they would be content with having a 'sea border' between Northern Ireland and Britain after *Brexit* (Hughes, 2017).

And for the Republic? The lack of an ultimate agreement would damage the Irish economy significantly as well as posing serious questions for the state's self-image.

> It has long enjoyed the luxury of not having to choose between being part of the EU on the one hand and being closely intertwined – culturally and economically – with the Anglo-American world. But as these two spheres drift apart, Ireland risks being pulled asunder if it tries to stay with both. It will have to think of itself, politically and

psychologically, as a more European country. Which would be all very well if part of the island were not being forced to define itself as much less European. (O'Toole, 2017: 31)

Notes

(1) London, ranked fifth, hosted 153 conferences in that year.
(2) There are some parallels with the Gibraltar situation here: see Chapter 11.
(3) With the physical expansion of Dublin port, the first of a fleet of '*Brexit* busting' large short-sea roll-on/roll-off ferries for *Cobelfret* of Antwerp was christened in Dublin port in April 2018 with the view to being operated directly from Ireland to Belgium and the Netherlands and thereby circumvent the UK's impediments (Finn, 2018; Pope, 2018).
(4) Whereby Northern Ireland would effectively remain in the single market and the UK would share a customs territory, leaving it unable to pursue an independent trade policy.

10 Inconvenient Cross-border Mobilities II: Expatriate Citizens' Free Movement Rights

10.1 Citizens on the Move: Visitors and Expatriates

As an issue highlighted in debates on EU withdrawal terms and likely outcomes, migration for work and retirement has been a growing topic of interest across a range of social science disciplines. Issues of access to citizenship and its associated rights (e.g. Mindus, 2017) possess important implications for both labour and immigration policy (Netto & Craig, 2016). The practical significance for tourism of citizens' rights to cross-border mobility in Europe is essentially three-fold:

- providing much needed labour in the hospitality sector: in restaurants, bars and accommodation provision, and for field labour to help harvest the fruit and vegetables supplied to the tourism and hospitality sector;
- generating two-way VF (visiting friends) and VR (visiting relatives) mobility to and from the UK (see Box 10.1); and
- generating additional 'domestic' tourism in the host country.

The role of expatriates in tourism appears to be both an underestimated (Backer, 2012, 2015) and an under-researched area (Dutt *et al.*, 2016).

Sørenson and Nilsson (1999; reprinted in Roberts & Hall, 2001: 41) drew upon Fees' (1996) analysis of conflicts between the 'local born and bred' and 'incomers', to construct a resident-visitors typology (for rural areas), in order to demonstrate the complexity of recreation-related mobility motivations and to suggest that rather than a host/guest dichotomy there can be identified a continuum of residential and visiting roles (Roberts & Hall, 2001: 40–41). This typology can been modified to suggest that the evolving mobility status of an expatriate (expat) may encompass, at different points in their life cycle, several of the positions (here termed 'stages') located on that continuum (Box 10.2), but that, for example, *Brexit* disruption can intervene and modify the outcome.

Box 10.1 Visiting Friends and Visiting Relatives: conceptual and practical issues

Although many commentators (and statistics) tend to employ the generic term 'VFR', researchers have recognised for some time significant differences in the nature and behavioural characteristics between visiting friends (VF) and visiting relatives (VR) mobilities.

Ramachandran (2006: 3) identified four 'divisions': VF, VR, visiting both F and R, and genealogy visits – visiting friends and relatives of the past.

Proportions of hosts receiving both friends and relatives appear to be relatively small – Seaton and Tie (2015) identified a level of about 18% – and tourists visiting both friends and relatives during the same trip may be proportionately low – Backer *et al.* (2017) found it applied to just 8% of international tourists.

Backer *et al.*'s (2017: 58) review of previous research (e.g. Lockyer & Ryan, 2007) showed that VRs outnumber VFs, often considerably; VRs generally stay longer and are more likely to include children in their number. Also VR hosts act as gatekeepers especially in providing information for their guests' visiting activities.

Dutt *et al.* (2016) found a relatively weak impact from (undifferentiated VFR) guests on expat-hosts' ability to learn about the destination in which they resided, such as opening hours and entrance fees.

Shani and Uriely's (2012) study of 'VFR' hosting identified both negative and positive impacts on hosts' quality of life: loss of privacy, extra expenditures, hard physical work and mental stress deriving from conflicting obligations of working and hosting. Such tensions may not, of course, apply to hosts who have otherwise retired, do not have full-time employment or young families to cope with. By contrast, the gratification of being able to host could provide joy, a sense of pride, opportunities to 'have fun' and even the moral privilege of claiming reciprocity at some future date. The authors recognised two types of hosts: (i) self-oriented, entailing maintaining the normal course of daily life and/or becoming a tourist 'in one's own back yard'; and (ii) guest-oriented, focusing on in-home hospitality and/or serving as a local tourist guide for guests.

Janta and Christou (2019) provided a gendered perspective on hosting.

(See also Williams & Hall's (2002: 9) later idealised aggregate model of tourism-migration relationships).

It is suggested that the hypothetical potential expat of 'stage' 1 is sufficiently attracted by an area/country to return in at least one of the roles identified in 'stage' 2 (partaking in VF/VR may obviate the need for a 'stage' 1). This may encourage further commitment to the area/country in the pursuit of a role identified in 'stage' 3, at which point the

Box 10.2 A residents-visitors-expats continuum

'Stage' 1 Potential expats

Holiday tourists: holiday tourists staying in the area.

Transit tourists: holiday tourists who consume the experiences and recreational values of the area, but are accommodated outside of the area.

'Stage' 2 Incipient expats

Day visitors: visitors having their permanent residence within day trip distance, using the area for recreation.

Staying with friends and relatives: visitors with socially defined connections to the area.

'Stage' 3 Part-time expats ('swallows')

Weekend visitors: regular users of a residence who 'go native' at the weekend.

Permanent tourists: two-home residents who reside in the country on a part-time basis. Alternation between countries of residence may transcend any work/holiday distinction.

'Stage' 4 Permanent/long-term expats

Resident tourists: persons who have moved to the country for a range of recreation-related reasons. The country is perceived as a more attractive place for life and family. They are now committed solely to the area/country for long-term residence and associated recreational experiences. They may act as VF/VR hosts, thus potentially generating further cycles.

Other incoming permanent residents: persons whose residence relates to functional reasons such as employment, but which may be tourism-related. Aesthetic qualities of the area may be less important.

'Stage' 5 Second generation; no longer expats?

Local born and raised residents: persons born and raised in the area. Not necessarily residing there always, their stays outside the area have a temporary character.

Source: Modified from: Sørenson and Nilsson, 1999: 8–9.

expat label may be adopted, at least in a part-time mode. A permanent and main or only home in the area/country is the characteristic of expat 'stage' 4, while 'stage' 5 takes this further into a second generation who may no longer associate themselves primarily with their parents' home country.

Of course, this process would be reinforced if the first generation expatriates sought and attained citizenship of their adopted country. Until that point, *Brexit* disruption could interrupt the process at any stage, on the one hand perhaps requiring the expat(s) to return to their country of origin or, on the other, convincing them that permanent absence and possible adoption of new citizenship was the appropriate lifestyle choice to make.

Box 10.3 EU free movement rights

1. All EU citizens have a right to reside in another EU member state for up to three months without any conditions other than the requirement to hold a valid identity card or passport. After three months certain conditions apply, depending on the status of the EU citizen (e.g. whether they are a worker or a student).
2. Those who opt to exercise their free movement rights are protected against discrimination in employment on the grounds of nationality. Provisions coordinating social security rules ensure citizens do not lose entitlements by working elsewhere.
3. EU citizens who have resided legally for a continuous period of five years in another EU member state automatically acquire the right to permanent residence there. To qualify for permanent residence, students and the self-sufficient must possess comprehensive sickness insurance cover throughout the five year period.

Source: McGuinness & Hawkins, 2017.

10.2 UK Citizens in the EU

The free movement rights of citizens across the EU (Box 10.3) has contributed to frequent travel between member states and the decision by many UK citizens to take up property in such countries as France, Italy, Spain and Portugal, either as their main or second home. There has been an imbalance between the EU expatriates in the UK and UK citizens in the EU, not only a 2:1 ratio in terms of numbers (Table 10.1) but also a difference in terms of life-styles. Most EU citizens in the UK are actively engaged in work, whereas between a third and a half of UK residents in the EU are retired.

While recognising statistical compilation inconsistencies between countries, at the time of the UK EU referendum there were between 1.2m and 1.3m UK nationals living elsewhere within the EU. In 2014, more than 8.7m visits were made by UK residents visiting friends and relatives (VFR) in other EU countries (Deloitte, 2016: 4). 'VFR' traffic would appear to have been important in sustaining the economic viability of a number of air routes. As the price of an air ticket has been reduced by low-cost airlines, EU citizens have been able to take more frequent trips and in some cases to commute between European countries for work, residence and/or leisure purposes. Higher air fares and fewer scheduled flights between the EU and UK were one predicted outcome of the UK's withdrawal from the EU (Deloitte, 2016: 6) (see Chapter 6), with serious implications for access to and from expat residential regions, and thus for mobility decisions.

An immediate impact of the referendum result for those dependent upon savings or pensions in sterling was that their cost of living rose

Table 10.1 UK-born residents/British citizens in other EU states and EU-born residents in the UK, 2015 (figures in thousands)

EU member state	UK-born residents	*British citizens	Country-born residents in the UK	EU member state	UK-born residents	*British citizens	Country-born residents in the UK
Austria	11	8.4	26	Italy	65	22.7	204
Belgium	27	25.0	32	Latvia	1	0.2	96
Bulgaria	5	2.6	77	Lithuania	3	0.2	147
Croatia	0.5	0.3	6	Luxembourg	7	5.5	1
Cyprus	41	24.1	31	Malta	12	6.7	20
Czech Rep.	5	5.2	42	Netherlands	50	41.4	79
Denmark	19	14.7	24	Poland	35	2.1	883
Estonia	0.5	0.3	18	Portugal	18	15.8	132
Finland	7	3.5	7	Romania	3	0.5	229
France	185	157.0	176	Slovakia	5	0.6	63
Germany	103	96.2	297	Slovenia	0.5	0.4	1
Greece	18	15.4	72	Spain	309	308.8	129
Hungary	7	2.6	87	Sweden	25	18.1	35
Ireland	255	112.1	411				
				Totals	1,217.5	890,299	3,325

Sources: Migration Watch, 2016; *ONS, 2017: Table 1.

following sterling's fall in value in relation to the euro. In the longer term, a UK exit from the EU could have seen the automatic right of UK citizens to live and work in EU countries coming to an end, or at least becoming more difficult until reciprocal agreements could be negotiated.

EU member states had two options in dealing with resident UK citizens:

- A declaratory system that would not involve Britons undertaking any extra paperwork, and would merely confirm that status was already held, whether as a permanent resident of more than five years or as a temporary resident with less than five years' residence.
- A constitutive system that required UK citizens to apply for new immigration status, mirroring the 'settled status' for EU citizens in the UK proposed by the Home Office (O'Carroll, 2018a).

Under international law the rights of UK citizens living and working in EU member states, and EU citizens living and working in the UK are guaranteed. Under the Vienna Convention on the Law of Treaties 1969, withdrawal from a treaty releases the parties from any future obligations to each other but does not affect any rights or obligations acquired under it before withdrawal. Further, any action to remove UK citizens currently in EU member states and any similar action against EU citizens in the UK would directly contravene Article 19 of the EU's Charter of Fundamental Rights, under which collective expulsions are prohibited (Migration Watch, 2016).

Figures published by Migration Watch (2016) derived from United Nations data, of UK-born residents in other EU states and of EU-born residents in the UK are shown in Table 10.1 alongside ONS (2017) figures for 'British citizens' (actually derived from Eurostat 2011 data). All of these figures should be treated with some caution, for reasons outlined in section 10.3.2 below and in, for example, McGuinness and Hawkins (2017: 5–6), who argued that:

- the *UK ONS Labour Force Survey* offered the best estimate of the EU migrant population living in the UK: 3.18m born in other EU countries, and around 3.16m nationals of other EU member states in 2015, while
- the *UN Global Migration Database* offered the best estimate for UK citizens elsewhere in the EU: 1.22m in 2015.

10.2.1 France

A special report by the ONS (2017) examining country-by-country migration patterns found that 43% of the UK citizens living in France in 2016 were aged between 15 and 54, while 46% were over 55 (with 19% over 65). Of those of working age, 52% were employed, 5.5% unemployed and 43% economically inactive. By contrast, more than half of the French citizens living in the UK were aged between 25 and 44. 65% of French people in the UK work in 'higher level professions': public administration, education, health and banking. By contrast, retirement appears to have been a driver for a significant proportion of the Britons in France: 67,000 draw a state pension, more than 40% of the total number. The attraction of France for expats replicates patterns of more conventional outbound tourism, with France being the second most popular destination for UK tourists after Spain (O'Carroll, 2017a).

10.2.2 Spain

Spain has been host to the largest number of UK citizens living in the EU (308,805) just under a third of whom (101,045) were aged 65 and over (ONS, 2017). However, when including 'swallows' (who spend half the year in Spain and half in the UK or elsewhere) and those UK residents who have not registered, according to a staff member of *Costa Blanca News*, the total number could be as high as 800,000 (Curran, 2018), or even 1 million full- and part-time residents (Finch, 2010). By contrast, the Spanish Statistical Institute (INE, 2018) has estimated that 102,500 Spaniards have been living in the UK (compare to Table 10.1 above).

Thus, unusually in Europe, Spain's balance of migration has been 'favourable' to the UK.

The Spanish foreign minister, Alfonso Dastis, suggested that Britons living in Spain would be allowed to stay there even if the UK left the EU without settling a withdrawal agreement. The fact that the Spanish economy benefitted from the 16m Britons travelling to Spain every year for holidays and visiting friends and relatives was clearly significant (Press Association, 2017), contributing more than 20% of total tourism income. The UK is the fourth largest market for Spain's goods and services, accounting for 7% of the total, and Spain's large services surplus is mostly the result of tourism.

> ... no country has a greater interest in the softest of Brexits. [than Spain]
> (Garicano, 2016: 128)

Nonetheless, with so many expatriates and second home owners being older people perhaps considering returning in their later years to their families (Gustafson, 2008) back in the UK, the uncertainty that *Brexit* posed could encourage significant numbers to bring forward their intentions. This would pose significant economic and social questions both for Spain and for the UK's national health service (NHS) (Sedgley *et al.*, 2017).

10.2.3 Expat representation

In October 2017 10 groups representing UK citizens in Europe (Box 10.4) sent an 'alternative white paper' to MPs and peers demanding their rights be prioritised in the UK's EU exit talks and guaranteed in the divorce agreement. They urged that UK citizens resident in the EU continue to have the same rights as they did before *Brexit*. Such rights should not be confined to residence but should include the continued rights to acquire citizenship, study, have academic and professional qualifications recognised, work, run a business, move freely between EU member states, and receive healthcare, pensions and other social benefits.

In the event of no agreement being reached over citizens' rights, expats' situations would be covered by third country legislation within the EU *acquis*. The perceived obfuscation and likely complications to arise from this were encapsulated in a *britishineurope.org* newsletter.

> This is intricately complex and, depending on your status, roams over a dozen directives. Not all of us though, will have rights under that legislation and, to complicate matters further, competence is shared with Member States and legislation can apply in parallel. This means that UKinEU citizens in different countries will face different rules and two member states, Denmark and Ireland, are wholly or partially outside EU common immigration policy. (BiE, 2018a: 3)

Box 10.4 EU citizens' rights pressure groups

> **British in Europe**: largest coalition of UK citizens living and working in Europe, with 10 full member groups and 12 associate groups ranging across the majority of EU countries, claiming to 'stand up for the rights of UK citizens in the EU and EU citizens in the UK', even if the name of the group might suggest otherwise.
>
> **ECREU – Expat Citizen Rights in EU**: a founding member of the Coalition of UK citizens groups in the EU: a lobby and self-help group, with an emphasis on supporting UK citizens in the EU, although its aims also include supporting EU citizens in the UK.
>
> **the3million** – representing the '3 million' EU citizens living in the UK, as a lobby group and support network, campaigning to safeguard and guarantee the rights of EU citizens in the UK and – albeit a clearly lower priority – UK citizens in Europe, after *Brexit*.
>
> **EERC – East European Resource Centre**: based in London, a lobby, information and support group for people from Central and Eastern Europe, encouraging integration into UK life.

Sources: BiE, 2018b; ECREU, 2018; the3million, 2017; EERC, 2015.

Brexit could bring an end to the European Health Insurance Card (EHIC) (see Chapters 3, 6 and 7) and to shared tax laws that had benefitted many UK home owners and expatriates living in the EU. With membership of the EU, the NHS had reimbursed UK pensioners for treatment in another country (the 'S1' benefit[1]). This agreement was put at risk when the UK government failed to offer to guarantee the continuation of reciprocal healthcare. Without reciprocal arrangements, UK expats would need to take out private healthcare insurance, which could be prohibitively expensive. The financing of health care for UK citizens in the EU – and vice versa – was clearly threatened (Fahy *et al.*, 2017).

Hitherto, living in Spain, for example, had been much cheaper than the UK. Campaigners warned that 'hundreds of thousands' of pensioners, particularly those with few or no savings, might need to return to the UK, if only temporarily (see Box 10.5), to be able to use NHS services unless reciprocal care arrangements remained in place. The existing system saved the NHS about £450m a year, according to a senior official at the Department of Health when questioned by the House of Commons EU withdrawal select committee. Spain charged an average of €3,500 a pensioner treated compared with €5,000 charged by the NHS (O'Carroll, 2017a).

Particularly vulnerable to changing circumstances were people who had retired to a warmer climate for health reasons (e.g. see Hardill *et al.*, 2005).

Box 10.5 Diasporic return migration as tourism

For returning expatriates, while 'home' is a real place connected with personal memories based on lived experiences, it may be rendered complex through the migration, prolonged absence and invariable comparisons with the host country.

The encounters and experiences afforded by the return visit may generate the emergence of a new consciousness in the returning expatriate about themselves, their family and their identities as migrants. This can result in a different (and differently perceived) relationship not only to the old home, but also to their adopted place of residence, drawing on the migrants' biographical past, personal heritage and legacy, as a key component of identity.

The contribution of diaspora studies can highlight the role of tourism in the relationship between home and host country, and the core of expat identity.

More research into the intersecting dynamics of tourism and migration is required, in terms of their influence on the subjective experiences and sense of identity of individuals, families and communities, and the wider consequences of these. How are they transformed in the longer term by such intrusive exogenous factors such as *Brexit*?

Sources: Huang *et al.*, 2016; Marschall, 2015, 2017; Mueller, 2015; Pearce, 2012.

ECREU argued that many of these expats, if they were forced to return to the UK, would be worse off financially and be a drain on the NHS (and perhaps the housing market).

10.2.4 Newest expats: *Brexiles*

As a counterpoint to UK citizens who have lived in EU member states for some time (and who might feel the need to return), one of the mobility consequences of the 2016 referendum result highlighted in the media was the choice of some British families to leave the UK and move to Europe for a perceived better lifestyle (*Brexiles*) (Lander & Tims, 2018; Moss, 2019).

Eurostat figures revealed that the number of Britons taking up citizenship of another EU country in 2016 – referendum year – increased by 165% from 2,478 in 2015 to 6,555, a figure more than four times higher than that of 2007. Germany was the biggest recipient in 2016 with 2,702 Britons taking citizenship there, more than four times the 2015 figure of 594 (Slawson & Crerar, 2018).

In 2018 183,993 Irish passport applications were received from the UK, the Irish passport office requiring increased staff numbers to cope with demand. In that last pre-*Brexit* year Northern Ireland experienced by far the UK's highest increase in house prices – 5.8% – possibly in part generated by demand from British buyers seeking eligibility for an Irish passport.

As the date for UK withdrawal approached, significantly increased sales of property to (hitherto) UK citizens with an eye to qualifying for EU residence were being reported from such regions as the Dordogne (Henley, 2018) and Costa Blanca (Curran, 2018).

Invited Contribution

10.3 Sustaining *La Dolce Vita*? An Expat View from Tuscany

Lesley Roberts[2]

This invited contribution by an experienced tourism academic and practitioner expresses the personal view of a UK expat living and working in northern Italy. Of course, no expat region is typical, nor indeed does any one individual expat face a typical experience. But the following piece usefully details some of the potential quotidian impediments for expats likely to result from Brexit, and raises pertinent questions for UK citizens to consider, both those in 'Europe' and in the UK.

10.3.1 Introduction

While for UK tourists the most popular foreign destination is Spain, a number have sustained a long-standing love affair with Italy, and with Tuscany in particular (Figures 10.1 and 10.2). Tourism is one of Italy's fastest growing and most profitable sectors, with around 4 million UK tourists visiting the country each year.

As home to the European Renaissance, Florence, in particular, is a key destination for students of art and art history, architecture, music, science, and philosophy, and a substantial part of the city's economy today is based on its cultural space and the tourism it generates (Lazzeretti, 2003). While Florence, capital of Tuscany, first became a popular destination on the Grand Tour from the 18th century, the romantic poets preferred the province's more remote mountains and coastlines, and Percy Shelley famously drowned off the coast of Viareggio, a cultural tie with the UK still recognised today. Tuscany's enchantment has continued to lure writers and poets from across the world, and tourists have followed, seduced by an irresistible blend of landscape, food, climate, and culture. Many of these tourists return to stay.

Figure 10.1 Expat residential territory in the hills of Tuscany

Figure 10.2 The walls and hills of Lucca, Tuscany

10.3.2 The expat community and the UK EU referendum

It would be hazardous to make anything other than an informed guess as to how many UK citizens have made Italy their home. There are a number of reasons for this difficulty. Different organisations treat the terms 'citizens' and 'residents' differently for data collection purposes. Many people do not register their residency in the country in which they are living, preferring to remain 'under the radar', and people move around more frequently than they can be counted (although one imagines this will no longer be possible for UK citizens following *Brexit*). As there have been no visa requirements between the EU member states, and no need to record UK citizens' arrivals and departures, data on the total number of British employed in the rest of the EU are not collected regularly. Dated figures published by the UK Office of National Statistics currently available are based on *Eurostat* data from 2010–11 national censuses (ONS, 2017). Using the European Council's figures, an estimate of the numbers of expats living in Italy suggests a figure as low as 37,000 (ECFR, 2016) (compare with Table 10.1).

It would be fair to say that the outcome of the June 2016 referendum stunned most of Britain's expat community in Italy. Its reaction in Florence, as reported by *The Florentine* (Farrell, 2016) was 'an indescribable cocktail of fear, shame, anger, shock and helplessness'. Expat communities may think of themselves as citizens of a European space or community, connected both to the place they have chosen to live and to their native countries. As a consequence, in addition to the higher-profile economic concerns raised by *Brexit*, expats may harbour a crisis of identity that should not be dismissed lightly (Farrell, 2016).

While, arguably, many people would not particularly mourn the fact, expats were perhaps the first casualties of *Brexit*, losing income almost overnight to the significant fall in the value of sterling immediately following the referendum (Collinson & Jones, 2016). The fall of over 10% against the euro was largely reported as a problem for holidaymakers, and this focus rather overshadowed its perceived impacts for the 1.2–1.3m UK expats living across the EU. Concerns over the status and rights of both UK nationals in the EU and EU nationals in the UK have been at the centre of halting *Brexit* negotiations since the outset, creating a disturbing sense of dislocation and insecurity.

10.3.3 *La Dolce Vita?*

Over half of expats of all nationalities claim to be in Italy for reasons of love or adventure (InterNations, 2017). But whatever precipitated decisions to relocate here, for British citizens, movement was facilitated by the UK's membership of the European Union. The lack of current residency data speaks volumes for the complexities that will be faced when Britons, as third country nationals, find that they have to be re-registered, with their circumstances documented and controlled in much closer detail. And since the UK government appeared to back off from its 'have your cake and eat it' stance (Roberts, 2017), we have seen that not having our cake is, ironically, going to take more than a bit of digesting. New processes of *extraco-munitari* registration for existing residents, and potential citizenship applications, are likely to present something of a challenge for expats. In Italy – where only one government has lasted the full five-year term since 1945 and change is a constant – this is perhaps more the case than in most other EU member countries.

Qualifying British nationals may apply for Italian citizenship on the basis of their existing rights, an option for those who have been resident here for five years or more (including those denied a vote in the 2016 referendum on the grounds that they had been out of the UK for 15 years or more). This may be the only way to maintain a European citizenship that will grant free movement. Until the end of the transition period all British citizens in Italy, as elsewhere in the EU, expect to be able to maintain their full rights as EU citizens. The UK government's (2017b) EU exit paper argued that British citizens in the EU should not assume that rights acquired under the EU's free movement rules would be guaranteed in the event of the UK's withdrawal. At the time of writing, however, applications for settled or pre-settled status were under way, allowing EU citizens in residence before UK withdrawal to continue to enjoy their rights to do so.

In January 2019 Italy became the first EU27 country to declare that UK nationals settled in the country would remain legal residents in the event of a 'no-deal' *Brexit*, following EU encouragement for all member states to do likewise (European Commission, 2018g, 2018h). In cases where eligibility for long-term residence had not been attained, citizens could remain until eligible to apply. In the event of a 'no-deal' exit, however, it was still expected that pensioners' rights to their index-linked pensions would be lost.

It has not helped that the rights of EU citizens living abroad have been largely unrepresented formally, both in their originating and host countries (but see Box 10.6), and many crucial questions have needed to be answered (Box 10.7). While the Italian Parliament is one of the few legislatures in the world to reserve seats for those citizens residing abroad, sadly, there is no such facility for British expats who have relied on the voluntary services of lobbying organisations such as *British in Europe* (https://britishinEurope.org) and *British in Italy* (www.britishinitaly.net).

Unlike the popular stereotype of the British expat as an entitled and sunburned hedonist, rarely moving from a deck chair, the smaller expat community in Italy represents curiosity, adventure, courage, and exploration – the characteristics and qualities that have underpinned a European (and global) mindset. Despite the challenges of an everyday bureaucracy for which Italy is renowned, expats support the economy, they work, set up businesses here and raise bi- or multilingual

Box 10.6 The *British in Italy* lobby group's claimed activities

- Lobbied the UK and Italian parliaments.
- A committee member gave evidence to the House of Commons Exiting the EU Committee in January 2017.
- Two other committee members gave evidence to the Italian Senate's joint committee investigating *Brexit* in July 2017.
- In conjunction with *Together Forward* and *the3million*, *British in Italy* had established good contacts with Italian politicians, including Sandro Gozi, *Sottosegretario* at the office of the *Presidente del Consiglio* with the Europe brief.
- *British in Italy* publications sent to Italian politicians.
- Through media contacts, word of mouth and a dedicated *Facebook* page regular contact was being made with British and Italian citizens affected by *Brexit*.
- Had meetings with the British embassy.

Source: British in Italy, 2017.

Box 10.7 Crucial expat questions requiring answers

- How will we have to apply for permission to stay?
- How is our right to work affected?
- What about subsidised university education?
- Will there be restrictions on property purchase?
- How will pensions be affected and will pensioners be entitled to healthcare?
- Will driving licences still be valid?
- How will the right to enforce a UK court judgement in Italy be upheld?
- Will the accompanied mobility of pets be affected?

And many more.

families. According to a financial complexity index created by The TMF Group (2018), Italy is the world's third most complex country in matters of accounting and tax compliance, after Turkey and Brazil. Further, at 43rd place out of 137 countries, Italy has been ranked as one of the least competitive countries in Western Europe and by far the lowest-scoring in the G7 (Edwards, 2017b).

Ask any expat and they will regale you with the stories of their company registration, taxation complicity, residency and citizenship applications, unsuccessful attempts to gain access to the much-prized *Tessera Sanitaria* (health card) over which, even without *Brexit*, there is much confusion, and regional rather than national agreements remain in place. There are also problems with utilities or postal services, to say nothing of attempts to navigate journeys using public transport or brave a post office. Those who attempt any of these without being able to speak Italian competently may be doomed to failure. Life *is* wonderful in Italy, truly lovely, but it is not for the faint-hearted. Thus, despite the challenges, and continued political uncertainty, expats remain happy with their lives here (InterNations, 2017). The advantages far outweigh the disadvantages for the bold with the ability to use the subjunctive.

10.3.4 The expat-tourism nexus

While at first glance there appears to be a paucity of research into the topic of expat travel, much falls under the umbrella of VFR tourism. A review of expat social media sites reveals extensive travel within Italy, with expats exploring the culture that drew them here, learning to recognise regions by their cuisine, architecture, dialects, and distinctive sociopolitical evolution. It appears to be a great pastime to eat one's way around the country, studying one of the birthplaces of western civilisation, and understanding the ebb and flow of ideas and their influences, or walking ancient pilgrim trails, such as the *Via Francigena* (the 'Italian Camino', actually extending from Canterbury to Rome) (Associazione Europea delle Vie Francigene, 2016). In fact, being so travelled and travel-savvy, expats even have their own targeted travel agencies.

Expats' patterns of travel differ from 'normal' visitor behaviour. Visits between second and new homes are frequent. Most expats have family, social and property responsibilities for which they may travel on a regular basis. Networks created by serial expats (those moving from one country to another) require more travel, creating and maintaining new international communities. In particular, such serial expats can exhibit extensive travel patterns, and Christmas, especially, sees people flying off around the world, often for a month or more. Indeed, international VFR tourism tends to reveal an above average length of stay compared to most other forms of tourism (Ramachandran, 2006) (see Box 10.1).

Expat tourism would appear to be a stimulus for a year-round VFR market in Italy. Anecdotal evidence from personal observation and experience might suggest that an expat family living in Tuscany can expect more than 12 visits a year from friends and relatives, with relatives staying for the longest periods – up to a month in some cases. Often such visits include a further holiday within the country, and visits by friends and relatives often provide the stimulus for host families to learn more about the country they are living in (Dutt *et al.*, 2016).

To expats, travel and tourism are not occasional occurrences; they are a way of life. It would therefore be an oversight to deny the contribution to a tourism economy made by a healthy expat community (even a small one), and dangerous to ignore the significance of its potential loss.

One of the major consequences of the UK decision to leave the EU, as with leaving the Customs Union for wider goods and services, will be increased administrative and documentation costs for UK travellers, and delays associated with additional bureaucracy. And because tourism is consumed at the point of production, additional factors, for example the future validity of UK driving licences, demonstrate further potential problems. The European Union's anticipated European Travel Information Initiative and Authorisation System (ETIAS) (likely to parallel the US ESTA scheme) (see Chapters 3 and 7) will offer further bureaucratic constraints, not only for holiday tourists but for expats and their networks of friends and extended families.

It was unlikely that citizens of a 'third country' such as the UK after *Brexit*, would enjoy a freedom of movement to live within Europe. Although those resident in Italy would probably be entitled to stay, they would be 'land-locked' in the sense that onward movement to other EU member states would no longer be automatic. Younger generations have been faster to appreciate the limitations this will place upon them. Perhaps the existing Schengen App will be matched by an EU App to allow British expats to be absolutely sure that they do not outstay their eligibility in EU countries – of particular concern to those who own their homes in the region.

As for the student expat, Italy boasts some of Europe's oldest and most respected universities with low tuition fees, yet only 4% of new entrants are overseas students (compared with an 11% average across OECD countries) (OECD, 2016). Language is a clear barrier (often, little English is spoken, even by young people), but many universities now offer programmes in English. Just as the north-east of England built its economy from the 1970s onwards on a student market, cities such as Milan, Bologna, Pisa and Florence are doing the same. From a tourism perspective, this visitor profile might appear ideal, supporting, as it does, cities' aims to promote the development of sustainable, responsible, cultural tourism in high-quality destinations.This is in line with the development of niche tourism that emphasises the unique selling points of specific destinations (OECD, 2016).

Cities such as Venice, Rome and Florence are at saturation point and are actively exploring ways to manage tourism behaviours. Today's tourists in Italy can be less discreet about their motives for travel than were their ancestors, and public displays of self-indulgence are frequently not a pretty sight, with the result that a number of

Italian cities, most recently Florence, have launched campaigns to help tourists to be more destination-aware (Firenzeturismo, 2018). Elsewhere in Italy, local politicians have intervened with comments on the behaviour of visitors, as in the town of Amatrice (northern Lazio, central Italy) where the mayor was moved to ask visitors not to take 'selfies' in front of mounds of earthquake rubble that represented lost communities, homes and lives following the August 2016 devastation there (Edwards, 2017a). By contrast, international students of art, art history, restoration, sculpture, renaissance architecture and philosophy – just some of the most studied topics in Florence – could be the tourists of choice; of perfect cultural exchange.

The Italian elections of March 2018 and May 2019 demonstrated a shift towards populist movements resulting from promises to the electorate that were unlikely to meet the demands of EU budgetary constraints. There was speculation that if a new government chose to uphold its promises to its supporters rather than to the EU, such a move would strike at the heart of the EU by compromising the Eurozone (Evans-Pritchard, 2018). But this fear was not new, and within Italy there was less of a focus in the media on Euroscepticism than had been the case in the foreign press. Rather, much more emphasis was placed on the 'fairer' taxation systems promised (Valentino, 2018).

For members of the expat community here in Italy, therefore, the elections and subsequent budgetary disagreements added further uncertainty to their lives. With the political shift, they had lost the parliamentary connections that expat lobbyists had built up over many years, and especially since the UK EU referendum. At worst, further economic strain would act as a disincentive to remain within the country, and any increase in nationalist tendencies would not sit comfortably with international mindsets.

10.3.5 Conclusion

We have to hope, therefore, that the centuries-long bonds between Italy and the UK will withstand whatever political compromise can be reached in pursuit of the professed 'will of the [British] people', and that meaningful connections can be sustained at personal and professional levels in order to maintain an outward-facing future for young and adventurous European citizens. It defies reason to restrict the movement of people within a European space when their futures are bound to a global economy. Truly, many outside the UK still view the country's economic focus and political isolationism with incomprehension. Those more sympathetic to a *Brexit* vote can also see that the changes required within the EU are more likely to be effected by struggles from within rather than by withdrawal and walking away.

It would appear that a healthy expat community has a stimulative influence on tourism both in terms of quantity and quality, generating VFR travel and additional spend. Study abroad for UK students should continue to be promoted (and EU programmes are available also to those outside of the Union), benefitting from cultural tourism, with its ability to inform and enlighten. And freedom of movement should be understood as a positive phenomenon, with its benefits promoted widely. For if the UK closes doors, it also closes minds.

Whether *Brexit* turns out to be a superb tactical manoeuvre on the part of the UK government, or a blind blunder in the dark, remained to be seen at the time of writing. Here in Italy, it is to be hoped that not everything that goes wrong is, in the words of comedian Francesco De Carlo *'tutta colpa della Brexit'*.

10.4 EU Citizens' Free Movement Rights in the UK

According to *People 1st* (2017b), other EU nationals comprised 39% of the UK tourism workforce and 48% of hospitality workers in 2016. As the UK unemployment rate had fallen, businesses had become increasing reliant on EU nationals to fill vacancies (see Chapter 6). This problem became critical in some regions and key cities such as London and in sectors such as hospitality and field agriculture. Migrant workers' rights and ability to stay in the UK had been critical to the functioning of many tourism enterprises (Tourism Alliance, 2017: 4).

The EU's position had been that, after the UK's withdrawal, all EU citizens living in the UK and all UK citizens living in the EU should continue to enjoy the same rights as before. But that would entail invoking the protection of the European Court of Justice (ECJ), which the UK government appeared to find unacceptable. Indeed, the UK government merely promised to 'seek to protect' EU citizens affected by *Brexit* in negotiations (O'Carroll, 2017a) (Box 10.8), as expressed in Lesley Roberts' piece above (section 10.3).

Figures released under a freedom of information request revealed that the UK Home Office was struggling to cope with the large number of applications submitted from EU citizens for permanent residency documents, with waiting times for some submissions increasing threefold. Permanent residency certificates – which could be issued to an EU citizen or their spouse during their initial five years in the UK – in late-2016 were taking an average of 116 days to obtain, according to Home Office figures. The average in 2015 was 43 days. Some 28,502 applications were submitted between July 2016 and June 2017, approaching double the 15,871 submissions in 2015–16 (Boffey & O'Carroll, 2017).

By volume, Poles topped the list of those seeking UK citizenship with almost 6,200 applications during 2017, an increase of 44%. The

Box 10.8 UK Government proposals (June 2017) on citizens' rights after *Brexit*

- All EU citizens present in the UK before a cut-off date and with five years continuous residence in the UK could apply for a new 'settled status' – akin to 'indefinite leave to remain' in UK immigration law. Subsequently, an application fee of £65 was set for this 'right'. It would be free for those who already had a valid indefinite leave to remain document or a permanent residence card.
- The cut-off date would be no earlier than 29 March 2017 (the date the Article 50 process was started) and no later than the date of the UK's exit from the EU.
- EU citizens in the UK before the cut-off date but without five years continuous residence in the UK could apply for a 'temporary status' in order to remain in the UK long enough to accumulate five years continuous residence, at which point they could apply for settled status.
- European leaders said the European Parliament would withhold its consent from any agreement that would treat EU citizens less favourably than hitherto.
- Campaign groups *the3million* and *British in Europe* warned of a lack of detail on how the government would protect the rights of UK citizens in the EU27.
- Polls conducted since the referendum had shown high levels of support for protecting the status of EU citizens lawfully resident in the UK.

Source: McGuinness and Hawkins, 2017.

sharpest rise in applications was from Germans, up from 797 the previous year to 2,338. The numbers of Italians applying increased from 1,109 to 2,950, and Spanish numbers almost trebled from about 500 to 1,400. A record 168,913 permanent residence documents were issued in 2017, two and a half times the previous year's 65,068. Nonetheless, compared to the overall numbers of EU citizens living in the UK, the number applying for citizenship remained small (Duncan & O'Carroll, 2018).[3]

An extra 1,200 staff were to be recruited by the Home Office over six months to establish an 'easy access' registration process for EU nationals. However, some of the largest employers in London, including hotel chains, supermarkets and banks were said to hold serious reservations over the UK government's capacity to establish such a system (Boffey & O'Carroll, 2017). Claims that the registration system had an error rate of 10% and a rejection rate of 27% anticipated such figures increasing dramatically when '3 million' people would have to apply for settled status (the3million, 2017) (see also section 6.7 above). Indeed, following the EU negotiations' 'flawed compromise' of December 2017, the UK Home Office was seen to be sustaining a 'hostile environment' policy (see Chapter 12). Hence the bold statement on the Home Office web pages relating to registration certificates and cards declaring that:

> If you already have a permanent residence document it won't be valid after the UK leaves the EU. (Home Office, 2018a, 2018b)

This would require over 3 million people to apply for the right to stay instead of being granted residence rights (the3million, 2017).

Following full implementation of the Home Office's new settlement scheme for EU and EEA nationals who had lived continuously in the UK for five years, more than 400,000 applications had been made by early April 2019 (Press Association, 2019).

10.4 Concluding?

The long period of uncertainty over the nature of the UK's withdrawal from and future relationship with the EU has been particularly difficult for expatriates both of UK origin in Europe and of European origin in the UK. Short- and longer-term mobility decisions revolving around the issues of leisure, health, employment and retirement have meant that expatriates' motivations and aspirations have varied widely and will, therefore, express different outcomes arising from impacts of the *Brexit* process.

In the absence of an agreement over the UK's EU withdrawal terms, the European Commission encouraged EU27 governments to take a 'generous' approach to protect the rights of resident UK citizens, and to consider granting temporary residence permits to allow time for people to secure long-term status (European Commission, 2018g, 2018h). The attitude of the French government was not untypical in arguing that it would guarantee the residence, employment and welfare rights of the 160,000 UK citizens living there, *provided* that the UK government offered the same guarantees to French citizens in the UK. At the time of writing, the UK government had not committed to such reciprocity.

Notes

(1) Formerly known as E106/E121, the S1 is a certificate of entitlement to health care in another EU/EEA country provided via the social security authority in the claimant's home country.

(2) Although the author studies Italian in a state school along with economic migrants and asylum seekers, her contribution here is restricted to those popularly referred to as 'expats' and who are British, this narrow definition including people who are overseas because of work commitments, those who have established businesses in international markets, as well as self-supporting retirees. This piece is in no way intended to under-estimate the contributions that can, and she hopes will, be made by those in Italy under different circumstances but who also seek equal opportunity to contribute to their host country.

(3) The Home Office resisted releasing such figures when first requested. It delayed release for eight months and only complied after an official warning from the Information Commissioner.

11 Inconvenient Cross-border Mobilities III: Gibraltar

> ... *the UK-Spain relationship is the picture perfect illustration of the economic and political benefits that the EU, and the Single Market, can bring about.*
>
> (Garicano, 2016: 126)

This chapter briefly evaluates Gibraltar's difficult position, having voted overwhelmingly to remain in the EU, in being largely dependent upon tourism and labour from across the Spanish border.

11.1 A Part

Gibraltar joined the EEC, later the EU, alongside the UK in 1973. It is one the UK's 14 overseas territories (UKOTs), of which nine have been associated with the EU via the Overseas Association Decision (OAD) adopted by the EU in 2013 (Clegg, 2016a). Some of these are major international tourism destinations, but each has benefitted from EU aid programmes disbursed under the International Development and Cooperation Directorate, amounting to at least €80m for the period 2014–2020 (Clegg, 2016b). Because of the UK's EU withdrawal, they would no longer be eligible when the next round of contracts began in 2021 (see also Chapter 12). Anguilla, the Turks and Caicos Islands and the British Virgin Islands in particular have needed significant support for reconstructing the infrastructure needed to support tourism following the devastation of Hurricane Irma in 2017 (Hare, 2017).

A number of UKOTs, such as Pitcairn (Clegg, 2017) have relied on EU funding to underpin environmental conservation measures laying the foundation for the island's tourism attractiveness (see Chapter 8). In addition to direct and indirect support for tourism, free access to the EU market has been beneficial to the Falkland Islands and Tristan da Cunha for their fisheries and agricultural exports, and Bermuda's financial services sector has benefitted from its close links with the EU, both in terms of the large market and the fact that the EU recognises Bermuda's regulatory system as equivalent to its own.

UKOTs have gained from free movement across the EU, to assist business links and educational opportunities. Recent and current projects have not only supported economies, but have helped UKOTs to address environmental challenges, not least climate change, disaster preparedness and the conservation of biodiversity. Deepening institutional links between the UKOTs and the EU, particularly via the European Commission, have provided more direct access to EU policy makers. Also, growing cooperation between the UKOTs and Dutch, French and Danish territories in the Overseas Countries and Territories Association (OCTA) has facilitated greater political visibility of such territories in Brussels and has generated a higher international profile generally (Clegg, 2017). By 2020 Gibraltar will have received €60m in EU funding since 1990 (House of Lords, 2017: 9).

Conceptually, Amoamo (2018) posited that *Brexit* as 'a substantial political shift' would invert or re-imagine peripheral islands – such as several of the UKOTs – as sites of reciprocal power projection, revising 'traditional' core-periphery relationships.

Gibraltar, however, has been the only UKOT actually part of the EU – under Article 355(3) of the Treaty on the Functioning of the EU – and, as such, the citizens of Gibraltar voted in the 2016 referendum (Clegg, 2016a). Aside from its geostrategic role, with a population of 28,000 within a three square mile isthmus (Figure 11.1), Gibraltar has exhibited 'intense social, political and economic activity, particularly in the financial and tourism sectors' (Mut Bosque, 2017: 1). This has reflected a transformation away from dependence upon the Royal Navy dockyard, such that access to the EU Single Market in services had become 'a fundamental tool' (House of Lords, 2017: 7).

The territory adopted a constitution in 2006 that confers a greater degree of autonomy than that obtaining, for example, in Scotland, and it guarantees Gibraltarians their right to self-determination should they wish it. Gibraltar's hybrid identity combines local cultural elements with a strong British consciousness. In 2002, an overwhelming majority rejected the proposal for a co-sovereignty arrangement between Spain and the UK (Garcia, 2002).

But in the 2016 referendum, Gibraltar voted by almost 96% to remain in the EU. There appeared to be three main reasons for this overwhelming support: (i) Gibraltar's workforce employs and depends upon both Spanish and foreign – mainly EU national – workers living in Spain; (ii) similarly, a large proportion of Gibraltar's tourists (albeit many of them British) also arrive through the land border from southern Spain; and (iii) the territory is heavily reliant upon imports.

All these factors require maintaining an open land border with Spain. The outcome of *Brexit* that would pose the greatest threat would be the

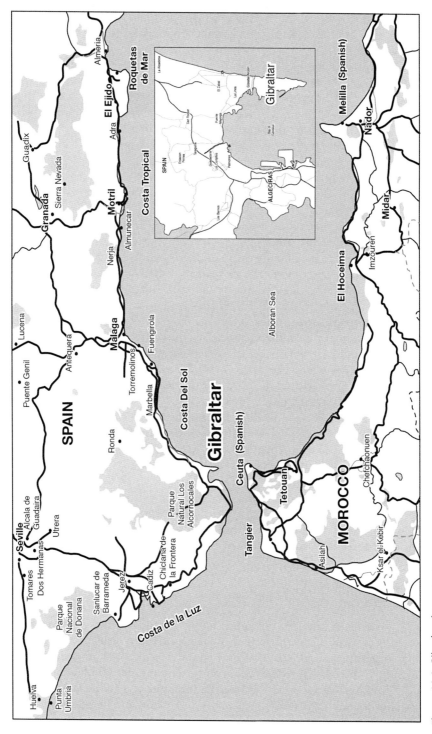

Figure 11.1 Gibraltar in context

complete closure of the frontier, which Gibraltar experienced between 1969 and 1985 as a result of Franco's attempts to isolate the territory (Macquisten, 2017).

Gibraltar now has an economy ranging from tourism, financial services, and e-commerce trading (including online gambling), to fuel bunkering and ship repairs. The economy employs 26,000 people, of which 12,000 commute across the frontier from Spain on a daily basis, 7,000 of whom are Spaniards. The remaining 5,000 are other EU nationals, including Danish, Dutch and French, who work in Gibraltar's tourism, e-commerce and financial services sectors (Chislett, 2017). Any change to the border control system would directly affect these workers' and tourists' accessibility, and ultimately, the economy of Gibraltar. It would also directly impact upon the neighbouring Spanish hinterland – the Campo de Gibraltar – encompassing the towns of La Linea, San Roque and Algeciras, areas of high unemployment (Jones, 2018a).

According to its Chamber of Commerce, Gibraltar generates around 25% of the surrounding Spanish area's GDP, making it the second-largest employer in the Campo de Gibraltar after the Spanish regional government there (Cañas, 2017). Spain's Andalucia region has one of the highest unemployment rates in Europe, at around 30%. With few prospects for increasing employment in the region, tightening controls at the frontier would appear to be self-defeating and politically risky for the Madrid government. The town of La Linea, in particular, was 'devastated' when Spain closed the border from 1969 until 1985, and locals fear that happening again (Jones, 2018b).

While both Spain and the United Kingdom have been EU member states, an open border at Gibraltar has functioned despite the UK not being in the Schengen Area. In 2010 the EU Commission had demanded that Spain 'fully respect EU law' by allowing free movement across the border, after the mayor of La Linea had disrupted frontier queues by threatening to impose a congestion fee on all those passing through. Normally, only non-EU citizens are subjected to documentation checks. However, a UK (and Gibraltar) exit from the EU risks such a free-flowing border becoming subject to Spanish government controls as an EU external border (Green, 2017).

11.2 Tourism

Tourism is one of the most important generators of revenue in Gibraltar. Although the territory has a modern cruise terminal for passenger liners and has developed a new international air terminal, the vast majority of visitors still enter via the land border (Table 11.1).

The tourism sector is vital to the functioning of Gibraltar: in 2017 it contributed £252m to the territory's economy, representing a 19% increase from the previous year. £176m of this (70%) was spent by

Table 11.1 Gibraltar tourism statistics, 2000–2018

Mode	Year						
	2000	2005	2010	2015	2016	2017	2018
Air passenger arrivals (actual numbers)	103,743	172,695	150,960⁺	218,720	270,067	277,784	186,231*
Land frontier visitor arrivals (millions)	7.031	7.434	11.071	9.626	9.464	9.877	8.764*
Land: arrivals by car (millions)	1.589	1.781	2.801	2.428	2.620	2.603	2.301*
Land: arrivals by coach (actual numbers)	14,763	9,805	8,174	6,461	6,931	7,111	5,025**
Sea: arrivals by yacht (actual numbers)	4,643	3,619	3,189	2,472	1,523	969	475***
Sea: cruise passenger arrivals (actual numbers)	–	–	–	342,942	404,005	404,994	394,530*
Sea: cruise crew arrivals (actual numbers)	–	–	–	145,847	166,559	167,497	163,994*
Total arrivals (millions, rounded)	7.14	7.61	11.23	9.85	10.31	10.73	9.51*
% of total arrivals by land	98	98	99	98	92	92	92

Key:
⁺The April 2010 figures were almost halved, presumably because of the disruption to air traffic over Europe caused by the Icelandic volcano ash cloud.
*First 10 months.
**First 11 months.
***First 9 months.
Sources: Government of Gibraltar, 2016b, 2017, 2018b.

excursionists from Spain. Some 92% of all *tourists* arrive over land (Government of Gibraltar, 2018a: 5; House of Lords, 2017: 8). If the EU's Schengen controls were to be strictly implemented, the tourism sector would suffer significantly, since the majority of Gibraltar's tourists travel from locations along the Costa del Sol on day trips.

Of course, with the new airport terminal the territory could encourage flights from additional destinations across the UK and elsewhere. Commercial air traffic to Gibraltar has increased considerably since 2013, when the new terminal opened, although Monarch Airline's demise following the referendum vote was not propitious. Royal Air Maroc Express introduced a service to Tangier and Casablanca in 2015. However, the numbers involved are still a fraction of those arriving by land.

While it would require a large number of additional air services to render visitor arrivals by air being anywhere near comparable to existing overland numbers, visitors arriving by air stay much longer and fill Gibraltar's expanding accommodation sector compared to the mostly day trippers arriving from the Costa del Sol. Thus any potential impediment to accessing European airspace would be of some concern, despite Spain's 2006 agreement to include Gibraltar airport in EU aviation measures (House of Lords, 2017: 8).

11.3 The Rock and a Hard Place

Following the 2016 UK EU referendum vote, the Spanish government initially resurrected its sovereignty claims. A joint sovereignty proposal was formally presented in September–October 2016 to the UN and the EU as an overall solution to Gibraltar's situation in relation to *Brexit*. The basis of this proposal was:

- transitional joint sovereignty between the UK and Spain, British and Spanish nationality, Statute of Autonomy (Article 144 of the Spanish Constitution);
- Spain would assume responsibility for external relations after the UK's effective withdrawal from the EU;
- Gibraltar would remain part of the EU;
- the border/fence and border controls would disappear.

The offer revived measures suggested in previous joint sovereignty proposals that had been firmly rejected by Gibraltarians in the 2002 referendum. Unlike the Blair/Aznar negotiations of 2001–2002 which led to that referendum, the UK government had adopted a harder stance since 2007, whereby it would not take any decision nor advance in any direction on sovereignty without prior Gibraltarian agreement. The UN General Assembly's Fourth Committee (Decolonization) reflected this UK position in its 2016 Gibraltar Decision. Gibraltar would boycott any move towards joint sovereignty (del Valle Gálvez, 2017): it is empowered to veto any change to its sovereignty by the territory's 2006 constitution (Ostrowski, 2019).

In stating a wish not to jeopardise the UK exit negotiations, Spain's foreign minister said subsequently that he would not subject any EU negotiation agreement to a desire for changing Gibraltar's status (Summers, 2017).

Nonetheless, Donald Tusk, the President of the European Council, generated considerable animosity when he announced in March 2017 that Spain would have a full veto on any deal concerning Gibraltar's status and sovereignty. This was unacceptable to most Gibraltarians, who argued that a president of one of the major EU institutions demonstrating a willingness to seemingly abandon the wishes of a community that largely voted to remain a part of the EU did not bode well for other member states.

This was followed in April 2017 by guidelines outlining the EU27's insistence that Gibraltar would be outside any future trade deal with the UK unless an agreement was reached in advance with Madrid over its future status.

In the March 2018 UK-EU agreement on the post-*Brexit* transition period, Gibraltar was to be included in the terms subject to bilateral agreement between the UK and Spain, with the latter wishing to retain an ultimate veto. Such an agreement was concluded in October 2018 as a protocol to be added to the UK's final EU withdrawal agreement (Stothard, 2018).

Unsurprisingly, Gibraltar's Chief Minister argued that a 'hard' *Brexit* would pose an 'existential threat' to the territory (McSmith, 2016). As early as September 2016 the Gibraltar government had delivered an economic impact assessment to the UK Department for Exiting the EU, emphasising the critical disadvantages Gibraltar could face in such a circumstance (Government of Gibraltar, 2016a).

Cross-party British support for protecting Gibraltar's sovereignty had been important to the people of Gibraltar, who despite voting over-whelmingly to remain in the EU, nonetheless were said to give greater value to being a part of the UK (Boffey & Jones, 2017).

On the other hand, the 2017 UK general election opened up new dimensions regarding the independence of Gibraltar's economy and tax status, with the possibility of the latter coming under closer scrutiny. This was significant as such status had been one of the primary incentives for financial services and e-commerce companies investing in Gibraltar. Subsequently, with 20% of UK motor insurance sold and perhaps 60% of UK online bets taken by Gibraltarian companies, the UK government agreed to guarantee access to the British market after *Brexit* for Gibraltar's financial services and online gambling sectors (Jones, 2018b).

11.4 Conclusion

The nature of the land border between Spain and Gibraltar is crucial for Gibraltar's tourism and travel sectors and for sustaining the territory's overall economy. The UK EU referendum brought the question of Gibraltar's sovereignty to the fore of Anglo-Spanish relations again, and for the first time since 2002 offered Spain an incentive to bring up the possibility of joint sovereignty.

That vitally significant sources of both tourists and labour continue to be able to travel across the border unhindered (and to also be able to arrive through European air space) must be a central consideration in Gibraltar's path after *Brexit*.

Section D

Global Britain?

12 Commonwealth

> *… 65 million of us do not come within a bull's roar of*
> *Britain's adjacent market of 450 million Europeans.*
> (Kevin Rudd, 2019: former Australian prime minister)

In light of some 'Leave' arguments suggesting that the UK's relationships with the Commonwealth could somehow supplant those with the European Union, this chapter addresses the triangular UK-Commonwealth-EU relationship and the place of tourism within it.

12.1 Commonwealth and the European Union

The Commonwealth has received relatively limited scholarly attention within the areas of mobility studies and political geography (Craggs, 2018; Dubow, 2017). Its 53 member countries extend across a fifth of the world's land surface, contain nearly a third of the world's population and produce around 15% of total world exports (Howell, 2016). Commonwealth countries range in size from India with a population of more than 1.3bn to the Pacific island states of Tuvalu and Nauru with 11,000 and 13,000 respectively; 32 contain populations of less than 1.5m.

Since at least 2010, some of those advocating the UK's removal from the EU had looked to the Commonwealth as a partial alternative framework, if only in terms of trade potential (Dominiczak, 2013; Hannan, 2016; Leave.EU, 2016; Nelson, 2011), despite the fact that the institution's power and influence had long been on the wane (Murphy, 2013, 2018; Onslow, 2015).

In its 2010 manifesto, the United Kingdom Independence Party (UKIP) had presented itself as 'the party of the Commonwealth', proposing to establish 'a Commonwealth Free Trade Area', which would account for 'more than 20%' of all international trade and investment and 'enable the UK to flourish outside the EU'. Although this argument was subsequently dropped, in its 2015 manifesto UKIP argued that the UK was part of a global community it called the 'Anglosphere', betraying a particular conception of the UK and *England*'s place in the world (Wellings & Baxendale, 2015: 136).

Not to be outdone, the Conservative Party manifesto for the 2015 general election pledged to 'further strengthen our ties with our close Commonwealth allies, Australia, Canada and New Zealand' (Conservative Party, 2015). And by the time of the EU referendum, at least one prominent 'Leaver' was happy to suggest that the UK had 'betrayed' the Commonwealth when it had joined the EEC in 1973, ignoring the fact that the majority of the Commonwealth did not agree with such a sentiment. What he was focusing on was the 'white' Commonwealth, where, indeed, there was some sense of betrayal (McIntyre, 2016) when the Commonwealth Preference System for the 'old dominions' was ended (May, 2013; Ward, 1997, 2001).

At the time of the 2016 referendum, however, the vast majority of Commonwealth members, and its secretary general, emphasised that membership of both organisations was compatible and mutually beneficial, and wished the UK to remain part of the EU (Howell, 2016; Marshall, 2016). This was not least because *Brexit* would throw the UK's future trading relationships into confusion. Existing EU agreements, which had allowed many developing Commonwealth countries access to the UK market on advantageous terms, were likely to lapse the moment *Brexit* took effect. It would be only after that point that the UK would have the legal capacity to negotiate trade agreements independently of the EU.

Far from being betrayed by the UK's 1973 accession, the majority of Commonwealth countries were 'accommodated' (Murray-Evans, 2016: 492): incorporated into the then EEC's system of preferential trade relations through the creation of the Africa, Caribbean and Pacific (ACP) group. The founders of the European project had favoured an EEC that promoted integration well beyond Europe (Hansen & Jonsson, 2011), even if it was born of a self-interested objective to construct a viable geopolitical power bloc and secure access to vital raw materials (Hansen & Jonsson, 2012).

> When it joined the EEC then, Britain did not so much betray the Commonwealth as secure the incorporation of the majority of its members into a set of broader European external economic relations ... As a result, the UK's trade relations with Commonwealth countries ... are now intricately intertwined with the EU's own system of external economic relations. In this sense, there is no straightforward choice between the EU and the Commonwealth. (Murray-Evans, 2016: 493, 490)

For many less-developed Commonwealth countries that had conducted a significant proportion of their trade with the UK, the prospect of UK EU withdrawal brought the risk of facing raised tariffs. The UK had been a major contributor to the EU aid budget, from which many Commonwealth countries had benefitted. Now, UK overseas dependencies faced the prospect of being denied access to this important source of development funding. Smaller Commonwealth countries also feared losing a powerful advocate within the EU, and would in future need to rely on Malta and Cyprus as their Commonwealth representatives in Brussels.

12.2 Two Commonwealths?

'Kith and kin' attitudes have a long pedigree within the British Conservative Party. Members of the party were instrumental in establishing the Anti-Common Market League (ACML) in 1961, a group which promoted British free trade interests and the maintenance of strong links with 'white' Commonwealth countries (Lloyd, 2016; Namusoke, 2016).

Before the 2016 EU referendum, the Commonwealth, or at least selected parts of it, was also being suggested as an alternative source of 'better-quality' immigrants. The implicit racism of such sentiments was given substance when in 2013 Boris Johnson, then London mayor, later to be foreign secretary and prime minister, proposed a 'bilateral free labour mobility zone' between the UK and Australia (Hewish, 2014).

The combination of the 'Leave' campaign's promises to reinforce ties with 'kith and kin' in Commonwealth countries with the xenophobia that defined the campaign, prompted the question for Namusoke (2016: 463) as to what the Commonwealth actually represented for contemporary Britain. She argued that the referendum campaign revealed two Commonwealths: one reflecting the backgrounds of UK ethnic minorities and one centred on the three majority white nations: Australia, Canada and New Zealand. Yet, as emphasised by a House of Commons briefing paper (Hawkins, 2016), reiterated by Namusoke (2016: 470),

> ... the majority of England and Wales' ethnic minority population have ties to the Commonwealth countries of India and Pakistan, and that the overwhelming majority of Commonwealth immigrants resident in the UK are nationals of non-white countries ... The perception of a Commonwealth in which Australians, New Zealanders and Canadians are the UK's closest cultural neighbours obscures the fact that, for decades, the Commonwealth in the UK has been represented by people of colour from the Asian subcontinent who have themselves become part of the fabric of British society.

And in the 2016 referendum, Britain's ethnic minorities voted by significant margins in favour of remaining within the EU (Ashcroft, 2016; Namusoke, 2016).

12.3 Tourism, Trade Myths and the Commonwealth

Looking at the responsiveness of Commonwealth tourism, remittances and aid to changes in UK economic conditions, Mitchell

Table 12.1 *Brexit* impact scenarios for Commonwealth tourism, remittances and aid

	Brexit impact scenarios					
	'Hard'		'Medium'		'Soft'	
Sector	Income effect	UK/US exchange rate effect	Income effect	UK/US exchange rate effect	Income effect	UK/US exchange rate effect
Tourism	0.025	0.012	0.028	0.013	0.031	0.015
Remittances	0.004	−0.001	0.005	−0.001	0.006	−0.002
Aid	0.130	0.023	0.152	0.024	0.130	0.024

Source: Mitchell *et al.*, 2018: 15.

et al. (2018: 15) suggested that, regardless of the outcome of UK-EU exit negotiations, *Brexit* was unlikely to have significant effects on Commonwealth countries (Table 12.1). This supported an analysis by the Commonwealth Secretariat (2016) which found that, because of the moderate share of UK contributions to Commonwealth countries' total inflows of tourism, remittances and aid, the aggregate impact from *Brexit* was likely to be 'contained'.

In their gravity model study, Mitchell *et al.* (2018) argued that Commonwealth tourism was mainly determined by distance, which was also a predictor of bilateral development flows. But distance played no role in determining the volume of Commonwealth remittances. The study did suggest, however, that increased trade between the UK and its Commonwealth partners could generate positive externalities, arguing that a 1% improvement in the value of total trade between two Commonwealth countries could increase tourist arrivals, remittances and aid by around 0.5% on average (Mitchell *et al.*, 2018: 16).

Brexit-motivated optimism over Commonwealth trade potential could be traced back to Katherine West (1994), who argued that the UK government's 'Eurocentric political rhetoric' had acted to disguise the importance to Britain of markets outside the EU. West (1994) noted the 'spectacular' economic growth of some Asian countries, suggesting that a 'myopic concentration' on Europe by UK policymakers impeded the exploitation of these markets.

Over the following two decades this notion was supported by such bodies as the Royal Commonwealth Society, the Commonwealth Business Council and the Commonwealth Secretariat. They identified what they saw as a 'Commonwealth effect' that rendered intra-Commonwealth trade cheaper and easier (Murphy, 2018a).

A Commonwealth Secretariat (2015) report claimed to prove the existence of this 'Commonwealth effect', with data suggesting that trade in goods and services between Commonwealth nations had been growing at almost 10% a year for two decades, and would 'exceed $1tn by 2020' (Commonwealth Secretariat, 2015: 40). The report claimed that when both bilateral partners were Commonwealth members, they traded on average 20% more, with, on average, 19% lower costs.

But, as the Commonwealth Secretariat also noted, there exist wide variations both in the extent to which individual Commonwealth countries trade with each other and individually with the rest of the world. This similarly applies to the movement of people. Average figures are, at best, misleading.

While some 44% of UK goods and services have gone to the EU annually, the entire Commonwealth has received around 9% of UK exports (Murphy, 2018). From the mid-1950s to the late-1980s, there was a steady reduction in the proportion of UK trade with the Commonwealth that reflected both a post-war revival of Western European economies and the 1948 General Agreement on Tariffs and Trade (the predecessor of the World Trade Organisation), which began to unwind the UK's preferential trading agreements with its remaining colonies and the Commonwealth.

UK withdrawal from the EU was unlikely to enhance commodity or services flows with Commonwealth countries. In the short term, it could actually act to reduce them while the UK attempted to replicate agreements previously in place under its EU membership. While Australia and New Zealand (and Canada) represented the 'acceptable' face of the Commonwealth to those 'Leave' supporters who continued to regard the citizens of the 'white' Commonwealth as 'kith and kin', Pacific Rim countries, especially through the APEC (Asia-Pacific Economic Cooperation) bloc had long taken the place once held by the UK market (McIntyre, 2016). Thus it was likely that a post-*Brexit* UK would need the Commonwealth a great deal more than the Commonwealth would need the UK, a reality that appeared to be largely reinforced by numbers and patterns of visitor flows enumerated below (Tables 12.2–12.5).

12.4 Tourism, Commonwealth and the EU

In terms of aggregate visitor numbers, the role of Commonwealth-UK mobilities would appear relatively small, certainly in comparison with the UK's tourism relationship with EU member states, being little more than an eighth of the volume.

Inbound and outbound visitor numbers between the UK and the Commonwealth are relatively modest and reasonably symmetrical in percentage terms at 11–14% of UK-EU levels and around 9% of UK totals (Table 12.2). In absolute terms, however, Commonwealth inbound numbers to the UK represent only around 55–60% of outbound, reflecting the wide range of Commonwealth destinations favoured by British tourists and VFR visits undertaken by UK residents with Commonwealth heritage.

Table 12.2 Basic and comparative Commonwealth visitor numbers, 2014–2018

Commonwealth and comparative inbound visitor numbers to the UK

Inbound visitor numbers to UK (m)	2014	2015	2016	2017	2018	Annual average % growth 2014–18
Total	34.377	36.115	37.609	39.214	37.905	2.5
From Commonwealth	3.072	3.244	3.386	3.705	3.499	3.8
From EU	23.009	24.213	25.486	25.586	24.795	1.9
% of all inbound visitors to UK						
From Commonwealth	8.9	9.0	9.0	9.5	9.2	–
From EU	66.9	67.0	67.8	65.2	65.4	–
Commonwealth inbound as % of EU visitors to UK	13.4	13.4	13.3	14.5	14.1	–

Outbound UK visitor numbers to the Commonwealth and other destinations

UK outbound visitor numbers (m)	2014	2015	2016	2017	2018	Annual average % growth 2014–18
Total	60.082	65.720	70.815	72.772	71.733	4.5
To Commonwealth	5.285	5.671	6.565	6.523	6.275	4.3
To EU	43.834	48.113	52.954	54.680	53.658	5.2
% of total UK outbound						
To Commonwealth	8.8	8.6	9.3	9.0	8.7	–
To EU	73.0	73.2	74.8	75.1	74.6	–
UK outbound visitor numbers to Commonwealth as % of UK outbound to EU	12.1	11.8	12.4	11.9	11.7	–

Sources: ONS, 2019b: Tables 2.10, 3.10; author's additional calculations[1]

As Table 12.3 reveals, however, sources and destinations differ markedly in levels of visitor activity. A clear VFR element is revealed in South Asian figures with outbound numbers being substantially greater than inbound by factors of between two and seven. Most other regions also reflect a heavy preponderance of outbound numbers, largely for 'holiday' tourism (Africa, Caribbean, and Mediterranean Europe).

How far were highly imbalanced visitor figures for the Caribbean impacted by the UK government's 'hostile environment' 'strategy'? (Box 12.1).

In contrast to other Commonwealth countries, Table 12.3 indicates that in 2018 for the 'old dominions' of Australia and New Zealand, inbound numbers were more than double outbound numbers from the UK, while roughly at a ratio of 8:5 for Canada.

In 2018 the level of spend of UK outbound visitors to Commonwealth countries was almost double the level of inbound spend in absolute terms (Table 12.4).

Table 12.3 UK-Commonwealth inbound and outbound comparative visitor numbers (in '000s), 2018

Commonwealth countries	Inbound numbers to UK	UK outbound numbers
Old Dominions		
Australia	1,003	356
Canada	850	543
New Zealand	216	98
South Asia		
India	511	980
Pakistan	102	572
Sri Lanka	20	140
West and East Africa		
Nigeria	107	138
'Other Africa'	169	757
South Africa	224	316
Caribbean		
Barbados	22	102
Jamaica	15	269
'Other Caribbean'	92	600
European Union		
Cyprus	96	745
Malta	74	659

Sources: ONS, 2019b: Tables 2.10, 3.10.

Overall, inbound Commonwealth spend was just over a third of EU totals and around just 15% of the total UK inbound spend. UK outbound Commonwealth spend was less than a quarter of that in the EU and 13–14% of the UK outbound total.

However, EU spend has been relatively low compared to the numbers involved, and in per capita terms Commonwealth spend levels in both directions were around double those for EU countries in 2018. Yet, Commonwealth per capita spend in the UK was less than two-thirds that of visitors from China.

As Table 12.5 indicates, and mirroring patterns of visitor numbers, the three 'old dominions' – Australia, New Zealand and Canada – exhibited higher inbound spending levels than UK outbound visitors to these countries in 2018. Nigeria also revealed a higher incoming spend. Other Commonwealth regions exhibited dominant UK outbound levels of spend: the Caribbean, South Africa and the EU Commonwealth states of Cyprus and Malta for 'holiday' travel and for South Asian VFR-related mobilities.

Box 12.1 'Hostile environment' and the Commonwealth

To meet the Conservative Party's pledged net immigration ceiling of 'tens of thousands', in 2012 then Home Secretary, Theresa May, aimed to create a 'hostile environment' for illegal immigrants, to make it much more difficult for them to lead an ordinary life in the UK (reinforced by the 2014 Immigration Act). Because of limited resources, this was in part about encouraging people to 'self-deport' by making their lives unpleasant.

Many former Commonwealth citizens were caught up in this process: what they had assumed were their unchallengeable rights to live normal lives in the UK appeared to have been removed. Those particularly affected were first generation Caribbean immigrants who arrived in the UK between 1948 and 1973 – the 'Windrush generation'[2] – as part of an open-ended invitation to settle in the UK, they had been given settlement rights and had not been required to obtain any specific documentation to prove these rights. Now, these citizens found themselves without the necessary paperwork to qualify for access to basic services, including citizenship.

Also as a consequence of the 'hostile environment', various UK cultural events were diminished because participants were denied visas. For the 2018 Edinburgh international book festival, for example, 'about a dozen' authors, mostly from the Middle East and Africa, had visa applications refused. At the *Womad* music festival at least three acts were unable to perform because of visa complications. For the same reason, some UK-organised conferences – particularly those development-related – were moving out of Britain to be convened in the near continent.

How could government reconcile promoting a 'GREAT Welcome' campaign for international visitors while pursuing such 'hostility', especially towards its own citizens?

Sources: Cain, 2018; Grant, 2019; Kirkup & Winnett, 2012; Savage, 2018b.

12.5 India

A UK government driven by the supposed anti-immigration rhetoric and actions of 'Leave' proponents presents a paradox for the non-white Commonwealth if the UK is seeking to expand (economic) relations with it. For example, in relation to the genuinely vast potential of the Indian market, many members of that country's policymaking elite were said to view the Commonwealth as little more than a quaint, somewhat odorous relic of British imperialism and patronage: 'Empire 2.0' (Hirsch, 2018).

While in 2018 inbound visitors from India numbered and spent between a half and two-thirds of UK citizens' outbound spend to India (Tables 12.3, 12.5), India has long desired a relaxation of UK visa

Table 12.4 Basic and comparative Commonwealth visitor spend statistics, 2014–2018

Commonwealth and comparative inbound visitor spending in the UK

Inbound visitor spend in UK (£bn)	2014	2015	2016	2017	2018	Annual average % growth 2014–18
Total	21.849	22.072	22.543	24.507	22.897	1.2
Commonwealth	3.307	3.264	3.320	3.635	3.417	0.1
EU	9.551	9.705	9.553	10.036	9.917	0.9
% of all inbound visitor spend in UK						
Commonwealth	15.1	14.8	14.7	14.8	14.9	–
EU	43.7	44.0	42.4	41.0	43.3	–
Inbound visitors' spend per capita (£)						
Commonwealth	1,076	1,006	981	981	977	–2.3
EU	410	396	386	390	400	–0.6
China	2,056	1,777	1,560	1,663	1,575	–5.8
Commonwealth inbound as % of EU inbound visitor spend in UK	34.6	33.6	34.8	36.2	36.2	–
Outbound UK visitor spend in the Commonwealth and EU						
UK outbound visitor spend (£bn)	2014	2015	2016	2017	2018	Annual average % growth 2014–18
Total	35.537	39.028	43.771	44.840	45.435	6.3
In Commonwealth	4.985	5.040	6.208	6.388	5.883	4.5
In EU	19.759	22.158	25.398	26.741	27.506	8.6
% of total UK outbound spend						
In Commonwealth	14.0	12.9	14.2	14.2	12.9	–
In EU	55.6	56.8	58.0	59.6	60.5	–
UK outbound visitor spend per capita (£)						
Total	591	594	618	616	633	1.8
In Commonwealth	943	889	946	979	938	–0.1
In EU	450	460	479	489	512	3.4
UK outbound visitor spend in Commonwealth as % of UK outbound spend in EU	25.2	22.7	24.4	23.9	21.4	–

Sources: ONS, 2019b: Tables 2.11, 2.13, 3.11, 3.13; author's additional calculations. See also Tables 7.9, 7.3.

restrictions on its nationals in return for trade liberalisation. Given the extent of UK citizens with South Asian heritage and the number of South Asian citizens with UK citizen relatives, rendering a sizeable potential V(F)R market, such a requirement poses no little paradox for those *Brexit* supporters looking to the Commonwealth for improved (i.e. self-interested trade) relations while also seeking to curb, albeit euphemistically, non-white immigration.

Table 12.5 UK-Commonwealth inbound and outbound comparative visitor spend, 2018

Commonwealth countries	Inbound to UK spend (£m)	UK outbound spend (£m)
Old Dominions		
Australia	1,044	575
Canada	676	530
New Zealand	255	178
South Asia		
India	491	702
Pakistan	110	383
Sri Lanka	33	130
West and East Africa		
Nigeria	152	127
'Other Africa'	191	776
South Africa	260	604
Caribbean		
Barbados	11	133
Jamaica	10	298
'Other Caribbean'	88	624
European Union		
Cyprus	60	478
Malta	36	345

Sources: ONS, 2019b: Tables 2.11, 3.11.

This issue overshadowed UK prime minister May's visit to India in November 2016, with accusations that the UK wanted India's business but not its people. At the April 2018 Commonwealth heads of state summit in London (theme: 'Our common future'), the Indian deputy high commissioner, reiterated the opinion that Commonwealth countries such as his would only undertake trade agreements if they included the removal of the restrictions on the free flow of people, a concept which many UK citizens who voted 'Leave' in 2016 appeared to be specifically voting against (Goodwin & Milazzo, 2017; Meleady *et al.*, 2017).

In conjoining concepts of Commonwealth and *Brexit* an internal contradiction is therefore presented; likewise the 'Leave' campaign's emphasis on 'Taking back control' with a 'Global Britain'.

12.6 Conclusion

Employment of 'the Commonwealth' as some kind of economic and psychological surrogate for a lost European relationship is flawed in several respects. Specifically in terms of visitor flows, numbers to and from the Commonwealth represent less than a tenth of total UK flows, while EU inbound flows represent around two-thirds of the UK total and outbound almost three-quarters. However, both inbound and outbound per capita visitor spend for Commonwealth countries is more than double the level for EU member states, in part reflecting longer-haul journeys and longer stays.

While tourism *per se* rarely featured in debates linking *Brexit* and the Commonwealth, the patterns of flows of people and money generated by tourism-related activity, not least when compared to EU-related flows, are both a significant dimension in their own right as well as reflecting and illuminating other relationship dimensions.

Notes

(1) Because of an incomplete breakdown of individual country data by ONS, 'Commonwealth' figures have been calculated in relation to the following entries: Canada, Cyprus, Malta, South Africa, Nigeria, 'Other Africa' (which excludes North Africa), India, Pakistan, Sri Lanka, Barbados, Jamaica, 'Other Caribbean'. There are thus some non-Commonwealth countries' data included, while certain Commonwealth territories are excluded (notably Bangladesh, Malaysia, Singapore, Fiji).

(2) Named after the *Empire Windrush*, the first ship to be employed in this process.

13 Pursuing the Chinese Market: Symbol of a 'Global Britain'?

In the 43 years since UK accession, the world has changed fundamentally, and the role of the EEC/EU within it likewise. Intra-European developments now carry considerably less weight than before. Globalisation has shifted the centre of gravity eastwards.

(Marshall, 2016: 458–459)

From long before the EU referendum, the UK government, tourism sector and higher education had recognised the emergent global significance of the Chinese outbound market, not least because of its relatively high spending potential. Unsurprisingly, an increasing emphasis on attracting Chinese visitors has been noticeable in recent years both in the UK and across Europe. This chapter therefore addresses the UK's engagement with that market and its post-*Brexit* significance for a 'Global Britain'. As Rong Huang warns in her invited contribution on Chinese students as tourists (section 13.5), it would be in nobody's interests for these UK sectors to become over-dependent upon one particular market, culturally, economically or geopolitically.

13.1 Market Characteristics

Seven years after the end of the 'Cultural revolution', China's outbound tourism journey was officially inaugurated in 1984, and by 2011 there were 70.25m outbound tourist trips, growing to 85m by 2016, spending $US 261.1bn (21% of the global total), although some 70% of (mainland) Chinese outbound tourists travelled to short-haul destinations: Hong Kong, Macau[1] and Taiwan (Dai *et al.*, 2017; VisitBritain, 2017b).

The (potential of the) Chinese market had become by far the world's largest, and growing. Three months after the EU referendum, the UK

Table 13.1 Peoples' Republic of China (PRC): key travel indicators

Population (millions) (June 2019)	1,420 18.4% of world's total
Population median age (June 2019)	37.3
GDP per capita PPP (US$)	13,802
Annual average GDP growth over previous year (%)	6.7
Annual average GDP growth over previous decade (%)	9.0
Population in possession of a passport (%) (UK figure is 76%)	7.0 (99m)
Estimated/projected outbound trips	2017 145m 2018 154m 2030 >400m
Weekly aircraft departures between PRC and UK	59*
Aircraft seat capacity 2018/19 (millions)	1.19**
APD incurred departing the UK for PRC	£75
UK visa requirements for PRC visitors from 2016	Applications for a 6-month standard visa automatically upgraded to a 2-year multi-entry document at the same cost: £89.

Notes: *April 2018–June 2019 saw 31 routes opened between China and Western Europe.
**Exceeded by China-Germany 1.57m, China-France 1.40m.
Sources: Casey, 2019; Insights, 2018; VisitBritain, 2017b; Worldometers, 2019.

and China signed an agreement lifting the limit on direct flights per week from 40 to 100 in each direction and removing a rule that only six airports in each could offer direct flights to the other country.

Key dimensions of China's potential are indicated in Table 13.1.

In 2018 Chinese nationals represented just 1.9% of all inbound visitors to the UK, and the equivalent of 2.9% of EU visitor numbers. There were comparable numbers of UK visits to China (Table 13.2).

Although figures dipped in EU referendum year, 715,000 Chinese visitors (Table 13.2) spent £1,126m in the UK in 2018 (Table 13.3), with a per capita spend more than four times that for EU visitors to the UK and 70–90% higher than for Commonwealth visitors (Chapter 12). Nonetheless, while numbers of Chinese visitors continued to grow after referendum year, per capita spend in the UK fell significantly over the 2014–18 period, as indeed had UK visitor spend in China (with per capita levels only around half that of outbound Chinese spend in the UK) (Table 13.3). Nonetheless, it is the continued high level of per capita spend, both absolutely and in relation to other inbound markets (Table 13.3), that renders the Chinese market so attractive to policy makers. This was emphasised by 2018 figures that saw China rise from sixth to fourth most important source of visitor revenue in the UK (after USA, Germany and France) (Table 7.10).

Table 13.2 Basic and comparative Chinese visitor numbers (including Hong Kong), 2014–2018

Chinese and comparative inbound visitor numbers to the UK						
	2014	2015	2016	2017	2018	Annual average % growth 2014–18
Total inbound visitor numbers to UK (m)	34.377	36.115	37.609	39.214	37.905	4.5
China inbound visitor numbers to UK (m)	0.392	0.529	0.525	0.636	0.715	1.9
China inbound as % of total visitors to UK	1.1	1.5	1.4	1.6	1.9	–
China inbound as % of EU visitors to UK	1.7	2.2	2.1	2.5	2.9	–
Outbound UK visitor numbers to China						
	2014	2015	2016	2017	2018	Annual average % growth 2014–18
Total UK outbound visitor numbers (m)	60.082	65.720	70.815	72.772	71.733	4.5
UK outbound visitor numbers to China (m)	0.477	0.537	0.601	0.569	0.652	9.2
UK outbound visitor numbers to China as % of total UK outbound	0.8	0.8	0.8	0.8	0.9	–
UK outbound visitor numbers to China as % of UK outbound to EU	1.1	1.1	1.1	1.0	1.2	–

Sources: ONS, 2019b: Tables 2.10, 3.10; author's additional calculations. See also Tables 7.7, 7.2.

13.2 GREAT China Welcome

The GREAT China Welcome initiative 'to make Britain the most welcoming destination in Europe for Chinese visitors',[2] was launched in spring 2014 to 'showcase, promote and develop' how the UK tourism and travel sector could respond to the requirements of Chinese visitors (VisitBritain, 2013). It was intended to show

> ... how serious we are about making sure Britain is ahead of the competition when it comes to attracting Chinese businesses and tourists. We are determined to encourage more Chinese people to come to our shores, enjoy our culture, heritage, food, sport, shopping, countryside and music and invest in our country. (Miller in DCMS, 2013b: 1)

In other words, the UK government was willing to offer the world's potentially largest market *visitability* and *investibility*, but presumably not *livability*.

The China Welcome programme aimed to secure 650,000 Chinese visits a year by 2020 to generate £1.1bn annually (DCMS, 2013b), both overall objectives being achieved in 2018 (Tables 13.2, 13.3). The programme sought to encourage and identify tourism and travel businesses taking specific steps to cater for Chinese travellers. The campaign

Table 13.3 Basic and comparative Chinese visitor spend data (including Hong Kong), 2014–2018

Chinese and comparative inbound visitor spending in the UK						
	2014	2015	2016	2017	2018	Annual average % growth 2014–18
Total inbound visitor spend in UK (£bn)	21.849	22.072	22.543	24.507	22.897	1.2
China visitor spend in UK (£bn)	0.806	0.940	0.819	1.058	1.126	1.0
China inbound as % of total inbound visitor spend in UK	3.7	4.3	3.6	4.3	4.9	–
China inbound as % of EU inbound visitor spend in UK	8.4	9.7	8.6	10.5	11.3	–
China inbound visitor spend per capita (£)	2,056	1,777	1,560	1,663	1,575	−5.8
EU inbound visitor spend per capita (£)	410	396	386	390	400	−0.6
Commonwealth visitor spend per capita (£)	1,076	1,006	981	981	977	−2.3
Outbound UK visitor spend in China and other destinations						
	2014	2015	2016	2017	2018	Annual average % growth 2014–18
Total UK outbound visitor spend (£bn)	35.537	39.028	43.771	44.840	45.435	6.3
UK outbound visitor spend in China (£bn)	0.598	0.494	0.543	0.495	0.476	−5.1
UK outbound visitor spend in China as % of total UK outbound spend	1.7	1.3	1.2	1.1	1.0	–
UK outbound visitor spend in China as % of UK outbound spend in EU	3.0	2.2	2.1	1.9	1.7	–
UK outbound visitor spend in China per capita (£)	1,253	920	903	870	730	−10.4

Sources: ONS, 2019b: Tables 2.11, 2.13, 3.11, 3.13; author's additional calculations. See also Tables 7.9 and 7.3.

would work with stakeholders to share customer insights, raise cultural awareness and encourage the training of qualified Mandarin-speaking guides, and in 2016 the UK government launched a £10m Mandarin Excellence Programme aiming for five thousand young people to become fluent Mandarin speakers by 2020 (UCL, 2016; see also Busby, 2017). Partly as a consequence, GCE 'A' level Chinese language saw an 8.6% annual growth in examination entries in 2018: the only modern foreign language to increase enrolments (Wood & Busby, 2018).

Appropriate stakeholders would appear on The GREAT China Welcome Charter with their details being shared widely in the market via a GREAT branded Chinese language directory and online database (www.ukgcw.com/) (DCMS, 2016: 14). To qualify, stakeholders needed to:

- meet specific criteria such as having first-hand experience in welcoming Chinese visitors within the previous two years;
- employ Mandarin-speaking staff; and
- accept *UnionPay*, China's national bank card, or have literature or signage in Mandarin.

As an example of collaboration efforts to attract Chinese visitors, Birmingham Airport and Chinese tour operator *Caissa* were cited in official literature as collaborating in order to increase the number of Chinese visitors *Caissa* would bring into the UK through the airport (DCMS, 2013b).

While 400 stakeholders (DCMS, 2016) had signed up to the Charter by mid-2016, this represented a relatively low take-up and a strong geographical bias in relation to the number of potential businesses within the sector. London accounted for almost 36% of hospitality participants, Scotland for less than 7%; 93% of independent hotels on the Charter were in England; Wales and Northern Ireland had none. Of tourist attractions featured on the Charter, 28% were based in London, and almost 73% in England. Of the others, 17.5% were in Scotland, 8.8% in Wales and none in Northern Ireland.

Clearly, the GREAT China Welcome Charter would be more effective if there was a less uneven distribution of participants. A large majority of Chinese tourists visit London, but many also travel out to visit historic cities and towns and to appreciate varied landscapes and regional cultures. Overall, it seemed, the UK was failing to promote alternative places of interest for this as for many other visitor markets. One response to this was Scottish Enterprise's expanding a 'China Ready' initiative first devised in Edinburgh (Dick, 2019).

Meanwhile, in January 2018, the EU and China established a collaboration programme – EU Ready for China – for the travel and tourism sectors. The EU's goal was to achieve a 10% annual increase in Chinese visitors, worth €1bn, as part of a range of commercial and cultural exchanges. This included EU-China Tourism Year (ECTY), funded by the EU, which sought to assist travel and tourism sectors to attract and better welcome Chinese visitors (Insights, 2018). The UK, of course, would be excluded from these developments.

13.3 Marketing Issues

In efforts to encourage more Chinese nationals to visit, UK tourism and hospitality businesses needed to address significant constraints, which the *VisitBritain* strategy recognised.

In 2016 UK Visas and Immigration (UKVI) opened three new visa application centres in China, bringing the number in the country to 15. There was, however, an awareness of the need to improve the availability of visitor information through the UKVI website and in visa application centres (DCMS, 2016: 14).

Although some Chinese citizens speak and/or read English, many do not. A large majority will also have trouble with spoken pronunciation and understanding various UK dialects. The needs and behaviours of the average Chinese tourist are also frequently misunderstood: the xenophobic mindset encouraged by some pro-*Brexit* supporters has only exacerbated this. Such factors can create an unwelcoming atmosphere. They can also render planning a trip to the UK as an individual difficult, although there has been a significant increase in recent years of numbers travelling individually and for pleasure rather than for business (Reckless Agency, 2017).

The average Chinese tourist stays longer in the UK than most other international visitors. In 2018 visitors from China spent an average of 14.3 nights (down from a peak of 22.0 in 2014) (24.8 for those from Hong Kong in 2018), compared to an overseas visitor average of 7.0 and an EU visitor average of just 5.4 nights (ONS, 2019b: Table 2.12).

China's new 'consuming' class, born after 1980, makes up more than 40% of its population, and represents two-thirds of the country's outbound tourists. In 2016, half of all Chinese visitors to the UK were aged 25–44 (VisitBritain, 2017b).This relatively affluent middle-class younger generation has mostly had further education paid for by their parents, with whom many still live (section 13.5 below). As such, this group carries a significant amount of disposable income, further enhanced in the UK with the post-referendum fall in the value of sterling (Reckless Agency, 2017).

Such travellers are more likely than average to be making their first visit to the UK: only 30% of the leisure visitors from China departing the UK in 2015 (excluding British expats) were repeating a visit they had made within the past 10 years, while 73% of departing visitors claimed they were 'extremely likely' to recommend Britain for a holiday or short-break (VisitBritain, 2017b).

The visitor attractions that appeal to Chinese tourists include lavish historical sites such as palaces and country houses, and high-end goods shopping, seeking luxury brands for themselves and for friends and families back home (Reckless Agency, 2017).

Some 80% of Chinese tourists pre-plan travel activities via the internet and social media, highlighting the importance for UK tourism and travel stakeholders to establish an engaging online presence. But engaging potential Chinese tourists on social media has many obstacles. Popular UK social media channels such as *Twitter* and *Facebook* are banned in China, while *Google* 'has been busy developing a tame search engine that might be acceptable to the Chinese government' (Naughton, 2018: 25).

More than half of the online Chinese population use the search engine *Baidu*, whose optimisation (SEO)[3] operates differently to Western systems (Reckless Agency, 2017). UK-based companies may find that promoting their business via a dedicated website in China is expensive. Working with a specialist Chinese travel PR agency and collaborating with Chinese tour operators can assist a targeted digital campaign from the UK, and UK SMEs forming alliances can help overcome marketing entry costs. For example, Blenheim Palace, a noted place of interest for Chinese tourists, has shared its marketing efforts with Bicester Village shopping outlet (Reckless Agency, 2017).

Any website without a .cn domain name may incur a penalty and *Baidu* gives precedence to websites that are hosted within mainland China due to the strict internet censorship. *Baidu* also prefers website content to be written in Mandarin. Due to the level of differentiation, a UK-based company would therefore be encouraged to have a separate top-level domain as well as a separate SEO strategy.

With £40,000 funding from *VisitScotland*, nineteen businesses joined forces to deliver an online campaign to promote Edinburgh's history and culture on the *Weibo* (a micro-blogging site described as a Twitter-Facebook hybrid) and *WeChat* Chinese social media platforms. With over 222m monthly active users, *Weibo* offers the opportunity to communicate with a mass audience at relatively low cost.

However, when top-rated attractions in the UK's capital cities were analysed in relation to their presence on *Weibo*, they revealed a patchy use of the channel. While 80% of the top 10-rated attractions in London held a *Weibo* account, only two in Belfast did so (Reckless Agency, 2017).

13.4 After *Brexit*

Continued weakness of sterling attracted high numbers of Chinese visitors in 2017 and 2018 (Table 13.2), although the long-term forecast was less certain, and a declining per capita spending trend since 2014 was notable (Table 13.3). If the UK was unable to negotiate a competitive trade settlement with the EU, and the EU's Ready for China programme was enhanced, Chinese visitors could be further attracted to Italy (already the most favoured destination in Europe), France and Germany, where high-end luxury brands would likely be available at competitive prices (Reckless Agency, 2017).

Ethics rarely seem to have impacted upon tourism development processes, despite long-term academic and activist pleas. But should it not be a cause for concern that a post-*Brexit* UK government may be committing much effort and resources in attracting large numbers of relatively high spending visitors from a country that has sustained the death penalty, has an estimated 1.1m people, mostly Muslim Uighurs, incarcerated in at least 44 internment and 're-education camps' (Byler, 2019; Kuo, 2018; Tisdall, 2018), and is pursuing an aggressive expansionist policy in the South China Seas?

> ... outbound tourism development helps to enhance China's voice in the international field of tourism to promote [the] soft power of its culture and values. We should take advantage of the rapid development of the outbound tourism to enhance the Chinese voice in international affairs ... (Dai *et al.*, 2017: 257).

Note: Some of the foregoing observations may not reflect the views of the following invited contributor.

Invited Contribution

13.5 Implications of *Brexit* for Chinese Educational Tourists in the UK

Rong Huang

13.5.1 Introduction

This case study discusses and evaluates implications of *Brexit* for educational tourism in the UK, and for Chinese international students in particular. It is organised into three parts. First, it presents impacts of Chinese international students in the UK. Second, it discusses Chinese international students' mobility and the implications of *Brexit*. Finally, it explores post-*Brexit* options for UK universities to attract Chinese international students.

By conceptualising the international student experience in relation to different tourist experiences, as theorised in the existing tourism literature, and through her primary research of Chinese international students in the UK, Huang (2005, 2008) argued that international students are not just normal students, but are more like tourists. International students in the UK promote invaluable economic, societal and cultural benefits, and generate employment across the UK. The prospect of stagnating or declining numbers of international students as a result of *Brexit* is a major concern to higher education institutions (HEIs) (Universities UK, 2017).

13.5.2 Impacts of Chinese international students in the UK

The provision of higher education for international students has become an important source of income for Western universities (Brooks & Waters, 2011). Due to their reputation for 'high quality', British universities attract a significant number of international students every year. The United Kingdom is currently the second most popular destination for international students after the United States (Universities UK, 2017). Given China's population of approximately 1.3bn people, the Chinese market provides huge potential for UK universities. This is for two main reasons: the increasing disposable income of mainland Chinese people, and the relaxation of outbound travel restrictions by the Chinese government (Cai *et al.*, 2000).

According to the Higher Education Statistics Agency (HESA, 2018a), for the academic year 2016–17, 81% of students studying in higher education in the UK were native, 6% were from the rest of the EU, and 13% were from the rest of the world. The total number of non-UK students studying in the UK was 442,375, and the number of Chinese students far exceeded any other nationality (HESA, 2018b): almost one third of non-EU students in the UK being from China. Figure 13.1 indicates the number of Chinese students studying in the UK since 2012.

Based on their detailed scrutiny of HESA data, Prazeres & Findlay (2017) revealed that many countries showed a consistent pattern of decline in sending students to the UK over the four-year span of their study. They reported that the largest reductions in numbers from within the EU were among students from Greece, Ireland and Germany.

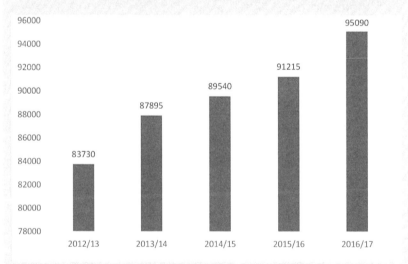

Figure 13.1 Number of Chinese students studying in the UK, 2012–2017.
Source: HESA, 2018b.

From outside the EU, there was also evidence of the UK becoming less attractive as a study destination. India, for many years, was a significant source of students to UK but has shown a serious collapse in student numbers (HESA, 2018b). Other non-EU sources (e.g. Pakistan and Saudi Arabia) also exhibited a reduction in international student migration to the UK over the four-year period. In the context of the highly marketised landscape of global higher education, Findlay *et al.* (2017) argued that it should be of great concern that the UK had become less attractive in so many key markets.

While the effects of the 2016 *Brexit* vote had yet to manifest themselves on incoming student numbers, Prazeres and Findlay (2017) suggested that changes to student visas, tuition fee increases and UK immigration policies had already contributed adversely to the profile of international student mobility to the UK (see also Chapter 7). It was anticipated that the immigration rhetoric associated with *Brexit* would have even more profound effects (Stone, 2016; Viña, 2016). According to HESA (2018b) data, China, as the biggest source country for UK higher education, was the only market showing a significant increase in student numbers (14% increase from 2012/13 to 2016/17). Students from China make up a sizeable share of international student fee income and their continued interest and contribution to UK higher education is critical to the UK's higher education sector (Hubble & Bolton, 2018).

Further, the importance of international students to the visiting friends and relatives (VFR) market is gradually being recognised by researchers (see Chapter 10), and several studies examining the student-related VFR market have been carried out in Australia, for example (Michael *et al.*, 2003; Taylor *et al.*, 2004; Tham Min-En, 2006; Weaver, 2003). Bischoff and Koenig-Lewis (2007) addressed this issue from a UK perspective and stressed that this segment can represent an important component of tourism demand and should receive greater marketing attention. According to *VisitBritain* (2017b), students made up 42% of the nights spent by Chinese visitors in the UK: China is thus a key market for study tourism.

Huang and Tian (2013) examined, from a tourism perspective, the experience of Chinese international students in the UK. Of their sample of 321, more than 42% had received visits from their families living in China since arriving in the UK to study. This provides strong evidence of the importance of Chinese international students for the VFR market in the UK.

13.5.3 Chinese international students' mobility and implications of *Brexit*

According to Packwood *et al.* (2015), international students enrolled at UK universities have been motivated by many factors;

the most frequently cited being the desire to enrol at a world class university (82%), and the opportunity to pursue an international career (61%). In relation to Chinese international students in the UK, Huang (2005, 2013) argued that the most important motivation factors for coming to the UK were to gain a better understanding of 'Western culture' through education, to gain a foreign qualification, and to travel. However, *Brexit* introduced a range of both negative and positive implications for Chinese international students' mobility to, and within, the UK.

Based on 2019/20 application figures, *Brexit* appeared not to be exerting a marked effect on international student recruitment (Morrison, 2019), with a 9% growth in undergraduate applications including a 19% increase in Chinese applicants (Adams, 2019a). By contrast, QS (2019) found that one in five prospective international students would be deterred from studying in the UK if the country left the EU.

Negative Implications

(a) The UK's unwelcoming image

Dennis (2016) argued that until the UK's EU withdrawal terms were clarified, international students, from Europe and the wider world, might view the UK as an unwelcoming country, possibly resulting in such students enrolling in other English-speaking countries. Viggo (2017) considered that 'visa and work restrictions combined with the global publicity that *Brexit* has received make the students feel less welcome in the UK'.

According to Hobsons (2017), when choosing a destination, international students are highly motivated by how welcome they feel. Giving international students a strong sense of welcome is therefore critical. However, most EU students (82%) considered the UK less attractive for study following the *Brexit* vote. Although some Chinese applicants were reported as not being influenced by the referendum outcome (Chen, 2016), others had considered alternative English-speaking countries as their study destinations for more open cultural possibilities. Emerging powers in Asia and elsewhere are fast becoming educational hubs and offer courses taught in English: *Brexit* may render these destinations more inviting to Chinese students.

Yet the UK's unwelcoming image has caused less concern than the Trump administration's tightening of visa requirements for Chinese students, especially for those studying science and technology, amid mounting concern about espionage (Turner, 2019). Hence the declining number of Chinese students applying to study in the US for the 2019/20 academic year.

(b) *Brexit* means immigration restrictions for European citizens and this might also mean more controls for non-EU visitors

The UK government's rejection of post-study work visa schemes (ComRes, 2016) taken along with the prevailing anti-immigration rhetoric of the UK's post-*Brexit* debate was likely to affect prospective students' perceptions and consideration of pursuing higher education in the UK (Universities UK, 2017). Some of these students could otherwise have been attracted to the UK as a destination for higher education, with the possibility of securing better employment opportunities upon graduation (Findlay *et al.*, 2017) than in their 'home' country (Bozionelos *et al.*, 2015). Kelly (2016) reported that such factors appeared to further dampen prospects of any recovery in international student numbers. The effects within EU countries were likely to be strongest, but perhaps more critical to UK universities would be the effect on perceptions of the UK as a study destination for students from China.

(c) Fewer EU students and more Chinese students mean even less international experience to be gained from the courses

Stone (2016) reported that too little was known of the likely status of EU students in the UK following Britain's exit. However, it seemed likely that fewer EU students would consider enrolling in higher education in the UK in the immediate future. China was already the largest contributor in terms of student numbers and revenue for UK higher education (HESA, 2018b), and continued growth in the number of Chinese students had been witnessed in many British universities in 2017 (Durnin, 2017).

Were there too many Chinese students in the UK? Chi (2018) reported that most Chinese students found they only mixed with their national counterparts in isolated groups on campus, and then often only with those on their own course, and thus gained limited international experience. Some Chinese parents feared that educational courses in the UK were losing their British character (Wei, 2017) as institutes made too many adaptations of the courses to recruit more Chinese students, and certainly some students reported that they had too many of their compatriots in their classes (Huang, 2013).

Positive Implications

While Highman (2017) clearly highlighted *Brexit*-related issues facing UK higher education, an analysis of articles published by the Chinese media outlets *China Daily* and *The South China Post* suggested that educational businesses and Chinese students focused on potential *Brexit* benefits and opportunities.

(a) A devalued pound means lower tuition fees …

According to Warrell (2017), non-EU international students in particular were able to make savings when the post-referendum weakened pound reduced the actual cost of their £13,500 annual undergraduate tuition fees. She reckoned that Chinese students would have benefitted from a 13% cut in the cost of fees between the June referendum and December 2016 (those students paying in US dollars would have saved 17%, and those using the South African rand 20%).

(b) … and cheaper shopping prices

After the *Brexit* vote, Loh (2016) reported that while many Britons were mourning the UK's imminent exit from the European Union, international students were celebrating the weaker pound which resulted in not only lower tuition fees but also cheaper shopping prices. Chinese media carried reports of students cheering lower tuition fees and going on shopping sprees in the UK (Chen, 2016; Phillips, 2016). Due to weaker sterling, more Chinese tourists were reported making purchases in London shops (Crabtree, 2017).

(c) Easier to find a job or start a business in the UK after graduation

A report from Learning without Borders (cited in Chen, 2016), a London-based global education platform, suggested that after the *Brexit* referendum it expected the number of EU students studying in the UK to be reduced. This meant less competition and more opportunities for Chinese students (Chen, 2016). Further, the UK government's proposal to reduce company taxes, combined with the fall in the value of the pound, made the UK more attractive for Chinese postgraduates who wanted to start a business there (Wang, 2017). Zhao (2017) also supported this argument, adding that post-*Brexit* Chinese students and investors could enjoy more equal status with those from EU countries, and therefore there could be more opportunities for them than previously.

13.5.4 Post-*Brexit* Options to Recruit More Chinese Students

There have been both short- and long-term implications of the *Brexit* vote for international student mobility. Considering the above discussion, the following should be noted by UK HEIs in addressing potential implications of *Brexit* for Chinese international students.

(a) Reconsidering motivations to study in the UK

Based on an extensive survey of international students, Hobsons (2017) concluded that they want a high-quality education at an affordable price in a place where they will feel welcome. Hence, when seeking to recruit international students, universities need to articulate their value proposition clearly and demonstrate what they can offer in each of these three areas.

Mok *et al.* (2017) reported that most Chinese students felt that their international learning experiences had positively contributed to their careers, citing career services and alumni networks as offering benefits. Soft skills – such as foreign language proficiency – were also reported as advantageous when applying for jobs. Chinese students and their parents are increasingly looking for British universities that can provide more rounded experiences, opportunities outside course study, and well-being support. Chi (2018) argued that those British universities which can create and demonstrate that they provide both 'soft skills' and effective welfare support for those that are thousands of miles from home, will quickly develop a positive and lasting reputation that can support more traditional selling factors.

Further, there is a range of activities which British universities could undertake to improve the employability of Chinese international students in the UK. For instance, as Huang *et al.* (2014) suggested, before universities emphasise the importance of employability to Chinese students, they should understand their students' views. In order to help Chinese students develop their employability, universities should consider using different types of assessments to examine student knowledge and abilities. The importance of the whole experience of being an international student in the UK should be emphasised. Further, based on apparent differences in perceived benefits of international education among Chinese students at different type of universities, Huang and Turner (2018) suggested that individual institutions might want to reconsider their strategies for enhancing employability and engaging with international students. In particular, individual institutions needed to appreciate that mainland Chinese students were not a homogeneous cohort (Iannelli & Huang, 2014).

(b) Keeping the UK as an attractive study destination

Corbett (2016) and Lowe (2016) reported that the UK had the most satisfied international students compared to its key competitors (e.g. USA, Canada, Australia and New Zealand) on all aspects of the student experience. Considering how they are treated on arrival, their learning experience, the support they receive and their living conditions, such students were more likely to recommend the UK system as a first choice. Further, a major attraction is that UK research has been conducted within an increasingly global network, and the UK has remained a world leader in research and innovation, boasting the highest field-weighted citation impact of any country, and ranking second in the world on the Global Innovation Index (OECD, 2013). Would such a status be threatened after UK withdrawal from the EU?

Rogelja (cited in Mardell, 2017) has warned against the 'commercialisation of British academia' and the 'transformation of higher education institutes into degree mills'. *Brexit* was threatening (or promising), to take UK higher education in a more China-orientated direction, and it was clear that careful consideration would be required concerning the likely consequences of pursuing an ever more China-orientated solution to potential *Brexit* problems (Mardell, 2017).

(c) Improving visa policies regarding international students

Restrictions on Chinese students wanting to work after their studies have damaged the attractiveness of Britain as a study destination. A critical review of the purpose and consequences of such policies is necessary to better develop visa requirements for international students.

As part of a package of government measures to boost numbers of overseas students after *Brexit*, international students were to be given visa extensions of up to a year to look for work in the UK (Adams, 2019b). But more could be done to help the UK achieve the objectives of its new international education strategy.

13.5.5 Conclusion

Following the earlier appraisal of Chinese outbound tourism for post-*Brexit* UK (sections 13.1–13.4), this case study has discussed the implications of *Brexit* for the recruitment of Chinese international students into UK universities, and their role as tourists and VFR tourism generators as numbers from other key international student source countries diminish. The Chinese media seem to have generated a positive discourse on *Brexit* for Chinese students. Potential options for UK universities to recruit more Chinese students emphasise the need to reconsider their motivations to study in the UK and to sustain the UK as an attractive study destination, not least through drawing on a diversity of source markets.

Notes

(1) As both Hong Kong and Macau have been functionally incorporated into the People's Republic (PRC) as 'special administrative regions', referring to them as outbound destinations is somewhat illusory. Of greater geopolitical significance is the PRC's employment of the propaganda term 'Greater China' to embrace both these territories and independent Taiwan.

(2) By 2017 at least Italy and France were still more popular than the UK.

(3) The process of affecting the online visibility of a website or web page in a web search engine's unpaid results.

14 Conclusion: The UK as a 'Great' Destination?

The British people have voted to leave the European Union. This creates new opportunities, from cutting red tape to forging new partnerships in emerging markets, as well as traditional strongholds like America. Working with the devolved administrations, our tourism bodies and the wider industry, we will seek to capitalise on this decision and show all [sic] our visitors how welcoming we are and that Britain truly is the home of amazing moments ... As ever, there will be challenges – from ensuring the industry has access to the right skills to security to connectivity. But, we will continue to work in partnership to make a success of Brexit and to ensure that this vibrant industry goes from strength to strength.

(DCMS, 2016: 15)

14.1 Retrospect

With the implications of the June 2016 EU referendum result continuing to unfold, this volume has attempted to make sense of the actual and likely consequences for tourism and related mobilities of the UK's EU exit. The evaluation began with a brief historical appraisal of the UK's relationship with 'Europe' (Chapter 1), highlighting the role of tourism in an increasingly interconnected continent. The domestic political context for the UK's 2016 EU referendum was addressed in Chapter 2, noting that the fall-out from the 'Leave' campaign's proven unlawful funding arrangements had yet to fully unravel.

In Chapter 3, a retrospective and prospective evaluation of the EU's approach to tourism was undertaken, specifically in the context of the UK's withdrawal. It was emphasised that the forging of a common European identity would continue, irrespective of the UK's position. A number of suggestions were offered in Chapter 4 for potential theoretical perspectives on the *Brexit*-tourism nexus, concluding that no one approach appeared to be adequate, given the complexity of EU withdrawal processes and consequences.

Critical observations on the UK government's early post-referendum *Tourism Action Plan*, promoted as a post-*Brexit* blueprint, were aired in

Chapter 5. Like the withdrawal proposal itself, the *Action Plan* left many questions unanswered, as did tourism sector appraisals of likely *Brexit* consequences. This examination led on to looking more closely at the wide-ranging implications for tourism supply and demand, in Chapters 6 and 7.

Longer-term implications of EU withdrawal and the ways in which tourism issues are inextricably enmeshed within broader interrelationships were explored in the following chapters. The ambivalence of the potential environmental implications of *Brexit* was perhaps overshadowed by the urgent need to mitigate climate change (Chapter 8). By contrast, Chapter 9, employing a mobilities focus, addressed the multilayered issue of the changing status of the border across the island of Ireland: a symbol of both EU withdrawal complexity and political irresponsibility. Much the same could be said of the mobility and rights issues of UK citizens resident in the EU27 and those of EU citizens living and working in the UK (Chapter 10), the prolonged uncertainty about which may only be resolved through a series of bilateral agreements that may not necessarily be consistent.

Although voting overwhelmingly to remain within the EU, not least because of a reliance on cross-border mobilities from Spain supporting its economy, Gibraltar was destined to leave the EU alongside the UK. The complications and potential geopolitical consequences of this were discussed in Chapter 11.

With one eye on the 'Global Britain' rhetoric of the 'Leave' campaign, Chapter 12 examined visitor mobilities and other factors contributing to the (putative) enhanced role of the Commonwealth following EU withdrawal. It identified the existence of more than one 'Commonwealth', and suggested that any emphasis on closer UK-Commonwealth ties was likely to be, at best, partial. Issues surrounding the perceived post-*Brexit* importance of the high spending, potentially enormous Chinese outbound market, were interrogated in Chapter 13, raising several questions concerning future policy (see section 14.3 below).

14.2 Frameworks for Further Work

How, from a tourism perspective, do we conceptualise the ongoing *Brexit* (hesitant) 'transition' (Chapter 4)? Is it a crisis or a series of crises through which and from which tourism can become more resilient? Does it represent some form of hybrid, perhaps negative mega-event, incorporating a soft-power meltdown and an exemplification of path dependency reversal? Or can it be viewed as a 'national' rebranding exercise (albeit one to which half the population does not subscribe), rippling its impacts across the British Isles, Europe, the Commonwealth, and beyond, reconfiguring environmental, economic, social, political and geopolitical policies and realities along the way? Perhaps it is all of these.

In terms of positivist approaches, there is clearly substantial longer-term potential for more closely researching *Brexit*-tourism inter-relationships, not least in the sometimes contradictory outcomes for outbound, inbound and domestic visitor dynamics (Chapters 6 and 7).

As multidisciplinary case studies, protection of citizens' rights (Chapter 10), the Gibraltar dilemma (Chapter 11), and, of course, the multifaceted issues attached to the border in Ireland (Chapter 9), offer myriad opportunities for detailed research into the relationships between mobilities and exogenous political impositions. And Ruth Craggs' (2014) research into 'hospitality diplomacy', not least within the Commonwealth, offers a platform for exploring the 'intimacy geopolitics' of decision making at several levels (Chapters 4 and 12).

14.3 The UK without the EU: A 'GREAT' Destination?

Tourism is emblematic of the EU's central concept of freedom of movement. The rhetoric from the UK government's early post-referendum *Tourism Action Plan* quoted above talked grandly in terms of the broad sunlit uplands that would emerge from *Brexit*. But in Chapter 4 of this volume it was posited that both the psychology and reality of the UK's exit from the EU were implicitly anti-tourism, not least in terms of the less than welcoming message that *Brexit* conveyed (sometimes explicitly), the expressions of a divided, embittered native population, and of the more constrained nature of individual (and services) mobility into and out of the UK that would follow.

If part of the *Brexiting* notion of 'Taking back control' was an implicitly anti-tourism rearguard action against the forces of globalisation (Chapters 2, 4), then a 'Global(ised) Britain' will confound that objective. As noted in Chapters 5 and 13, a key ingredient in the UK government's post-*Brexit Action Plan* for tourism strongly favours attracting the Chinese and other Asian tourist markets (and investment) that exhibit high per capita spending. As Chapter 13 noted, however, there are considerable human rights and geopolitical issues surrounding China's internal and external behaviour, which, in the longer term, may challenge the wisdom of the UK's investing so much effort and resources into attracting *renmimbi*.

This is not unique, given that the EU is also wooing high spenders from Cathay. But for an 'independent' UK, it does raise serious questions concerning the strategic need to diversify markets, a requirement noted, for example, in respect of the UK's higher education overseas student recruitment practices (Chapter 13).

While it may be too soon to evaluate how far brand *Brexit* can act to promote an appealing UK on the global stage of *visitability* and *investibility*, the government's continuing 'hostile environment' for (illegal) immigrants and some refugees, that has caused a great deal of collateral social and psychological damage (Chapter 12), has done little to enhance the country's apparently waning soft power reserves.

Domestic issues that were exploited by the 'Leave' referendum campaign – collective expressions of being overlooked and left behind in 'communities' with diminishing public and private services – have only been addressed subsequently with token gestures. Government resources have been diverted and so focused on EU withdrawal processes that other, notably pressing social welfare and environmental health issues, have been largely ignored, exacerbating the polarised nature of UK society (Chapters 2 and 4).

In sum, these issues do not add up to a 'Great' basis for welcoming visitors, nor for encouraging their return.

14.4 Reflexivity ... and Beyond

In the course of this volume, the author and invited contributors have attempted to be as dispassionate as is realistic for academics who themselves may have a great deal to lose from the UK's withdrawal from the European Union. Indeed, one of the questions that Bristow *et al.* (2016) posed concerns how, as an academic community studying and questioning many of the issues raised and unleashed by *Brexit*, we can critically consider our roles as researchers and educators in fostering constructive debate, challenging underlying racist, class, regional and other tensions, and dissipating the damaging effects and consequences of *Brexit*.

This volume has represented a small, but hopefully informed, contribution to the debate, notably of course, in terms of tourism-related mobilities. It has been written between the 2016 EU referendum and the final months before the UK's withdrawal, and published during the withdrawal transition period. Given the constraints that this timing has imposed, it is hoped that this volume provides a springboard for reflection on and analysis of the consequences of withdrawal on tourism-related mobilities to go 'wider and deeper'.

Much political uncertainty stalks Europe: never was there a better time to scrutinise tourism processes and to raise important questions concerning the changing geopolitical context within which they are undertaken and the challenge of confronting often cynical *Realpolitik*. Not least, with urgently pressing climate change/catastrophe and species extinction warnings often obscured by the *Brexitcentric* black hole of public debate, we need to be more critical in questioning the very nature and purpose of tourism travel itself.

References

Abbott, L. (2016) *Greenland: History and Culture, Tourism and Travel Guide*. Abidjan, Côte d'Ivoire: Sonit Education Academy.

ABTA (2017) *Making a Success of Brexit for Travel and Tourism*. London: ABTA. https://abta.com/tips-and-latest/reports-and-trends/making-a-success-of-brexit-for-travel-and-tourism (accessed 14 August 2018).

Adams, R. (2019a) Surge in applications from abroad to study at UK universities. *The Guardian* 7 February, p. 13.

Adams, R. (2019b) Visa extension to boost numbers of overseas students in UK after Brexit. *The Guardian* 16 March.www.theguardian.com/education/2019/mar/16/visa-extension-overseas-students-uk-brexit (accessed 20 April 2019).

Adonis, A. (2018) The ticking time bomb hidden in this Brexit bill. *The New European* 15 March, p. 11.

Agovino, M., Casaccia, M., Garofalo, A. and Marchesano, K. (2017) Tourism and disability in Italy. Limits and opportunities. *Tourism Management Perspectives* 23, 58–67.

AHDB (2017) *Brexit Scenarios: An Impact Assessment. Market Intelligence*. Stoneleigh Park, UK: Agriculture & Horticulture Development Board.

Alibhai-Brown, Y. (2017) Don't let leavers control the narrative any more … *The New European* 26 October, pp. 10–11.

Alogoskoufis, G. (2016) Greece and the upcoming Brexit negotiations. In C. Wyplosz (ed.) *What to do with the UK? EU Perspectives on Brexit* (pp. 53–60). London: CEPR Press.

Amin-Smith, N., Cribb, J. and Sibieta, L. (2017) Reforms to apprenticeship funding in England. In *Institute of Fiscal Studies Green Budget 2017*. London: IFS. www.ifs.org.uk/publications/8863 (accessed 23 December 2017).

Amoamo, M. (2018) More thoughts on core-periphery and tourism: Brexit and the UK Overseas Territories. *Tourism Recreation Research* 1–16. Doi:10.1080/02508281.2018.1455015

Anastasiadou, C. (2008a) Stakeholder perspectives on the European union tourism policy framework and their preferences on the type of involvement. *International Journal of Tourism Research* 10 (3), 221–235.

Anastasiadou, C. (2008b) Tourism interest groups in the EU policy arena: Characteristics, relationships and challenges. *Current Issues in Tourism* 11 (1), 24–62.

Anastasiadou, C. (2011) Promoting sustainability from above: Reflections on the influence of the European Union on tourism governance. *Policy Quarterly Journal* 7 (4), 27–33.

Andrew, D. (2016) The UK leaving the EU would remove funding sources for museums working internationally. *Museums and Heritage.com* 22 February. http://advisor.museumsandheritage.com/blogs/the-uk-leaving-the-eu-would-remove-funding-sources-for-museums-working-internationally/ (accessed 16 December 2017).

Anghel, S., Drachenberg, R. and Dinan, D. (2018) *From Rome to Sibiu, the European Council and the Future of Europe Debate*. Brussels: European Parliament Research Service. www.europarl.europa.eu/RegData/etudes/STUD/2018/615667/EPRS_STU(2018)615667_EN.pdf (accessed 20 December 2018).

Anon (2016) Concerned local anti-Brexit group formed. *Dundalk Democrat* (Republic of Ireland) 17 September. www.dundalkdemocrat.ie/news/home/216158/concerned-local-anti-brexit-group-formed.html(accessed 6 May 2018).

Anon (2017a) Blue is the colour. So what? *The Guardian* 23 December, p. 34.

Anon (2017b) Britain's disgraceful indifference to the Irish border could be lethal. *The Guardian* 20 November, p. 26.

Antonucci, L., Horvath, L., Kutiyski, Y. and Krouwel, A. (2017) The malaise of the squeezed middle: Challenging the narrative of the 'left behind' Brexiter. *Competition and Change* 21 (3), 211–229.

Arscott, J. (2017) Clouding our own horizons. *The New European* 14 December, p. 27.

Ashcroft, Lord (2016) How the United Kingdom voted on Thursday … and why. *Lord Ashcroft Polls*, 24 June. http://lordashcroftpolls.com/2016/06/how-the-united-kingdom-voted-and-why/ (accessed 5 August 2018).

Associazione Europea delle Vie Francigene (2016) *The Via Francigena*. Piacenza Italy: Associazione Europea delle Vie Francigene. www.viefrancigene.org/en/ (accessed 7 October 2018).

Asthana, A. (2017) MPs' fury over edited Brexit impact reports. *The Guardian* 28 November, pp. 1, 7.

Avraham, E. (2016) Destination marketing and image repair during tourism crises: The case of Egypt. *Journal of Hospitality and Tourism Management* 28, 41–48.

Aykin, S.M. (2012) A common tourism policy for the European Union: A historical perspective. In P.M. Burns (ed.) *Controversies in Tourism* (pp. 23–40). Wallingford: CABI.

Bâc, D.P. (2015) Endogenous challenges for the tourism industry. *Quaestus* (Timişoara, Romania) 6, 231–239.

Bachtler, J. (2017) Brexit and regional development in the UK: What future for regional policy after Structural Funds? In D. Bailey and L. Budd (eds) *The Political Economy of Brexit* (pp.127–156). London: Agenda Publishing.

Bachtler, J. and Begg, I. (2017) Cohesion policy after Brexit: The economic, social and institution al challenges. *Journal of Social Policy* 46 (4), 745–763.

Bachtler, J. and Begg, I. (2018) Beyond Brexit: Reshaping policies for regional development in Europe. *Papers in Regional Science* 97 (1), 151–170.

Bachtler, J., Mendez, C. and Wishlade, F. (2017) *Reshaping the EU Budget and Cohesion Policy: Carrying On, Doing Less, Doing More or Radical Redesign?* Glasgow: European Policies Research Centre, University of Strathclyde, EPRP 104.

Backer, E. (2012) VFR travel: It is underestimated. *Tourism Management* 33 (1), 74–79.

Backer, E. (2015) VFR travel: Its true dimensions. In E. Backer and B. King (eds) *VFR Travel Research: International Perspectives*. Bristol: Channel View Publications.

Backer, E., Leisch, F. and Dolnicar, S. (2017) Visiting friends or relatives? *Tourism Management* 60, 56–64

Balcombe, S. (2016) *Delivering World Class Product: The Discover England Fund*. London: VisitBritain/VisitEngland.

Barkham, P. (2018) EU farm rules blamed for insect and bird declines. *The Guardian* 24 March, p. 41.

Barrett, A., Bergin, A., Fitzgerald, J. *et al.* (2015) *Scoping the Possible Economic Implications of Brexit on Ireland*. Dublin: ESRI Research Series 48. www.csri.ie/pubs/RS48.pdf (accessed 5 December 2017).

Barrett, A. and Goggin, J. (2010) Returning to the question of a wage premium for returning migrants. National Institute Economic Review 213 (July)/Bonn: Institute for the Study of Labor (IZA) Discussion Paper 4736.

Barrett, A. and Morgenroth, E. (2016) Ireland and Brexit. In C. Wyplosz, (ed.) *What To Do With the UK? EU Perspectives on Brexit* (pp. 69–77). London: CEPR Press.

Bauman, Z. (2000) *Liquid Modernity*. Cambridge: Polity Press.

Bauman, Z. (2004) *Conversations with Benedetto Vecchi*. Cambridge: Polity Press.

Becker, S., Fetzer, T. and Novy, D. (2017) Who voted for Brexit? A comprehensive district-level analysis. *Economic Policy* 32 (92), 601–650.

Beerli, A., Díaz Meneses, G. and Moreno Gil, S. (2007) Self-congruity and destination choice. *Annals of Tourism Research* 34 (3), 571–587.

Benn, M. (2018) Feeling blue. *The New European* 22 March, p. 30.

Bennett, R. and Woolcock, N. (2018) A level results day 2018: Chinese more popular than German. *The Times* 16 August. www.thetimes.co.uk/article/a-level-results-day-2018-record-proportion-attending-university-ghl8dtkvp (accessed 18 August 2018).

Bergin, A., Garcia-Rodriguez, A., McInerney, N., Morgenroth, E. and Smith, D. (2016) Modelling the medium to long term potential macroeconomic impact of BREXIT on Ireland. *ESRI Working Paper* 548, Dublin.

BiE (British in Europe) (2018a) The Brexit cliff edge and our rights. *BiE Newsletter* February, p. 3. www.britishineurope.org/newsletter060218finmidres.pdf (accessed 24 February 2018).

BiE (British in Europe) (2018b) Who we are. https://britishineurope.org/who-we-are (accessed 24 February 2018).

Bischoff, E. E. and Koenig-Lewis, N. (2007) VFR tourism: The importance of university students as hosts. *International Journal of Tourism Research* 9, 465–484.

Bloom, J. and Hoggan, K. (2017) Apprenticeship numbers fall by 59% after levy imposed. *BBC News* 23 November. www.bbc.co.uk/news/business-42092171 (accessed 25 November 2017).

Boffey, D. and Jones, S. (2017) Gibraltar heading for an abrupt exit from single market, says Spain. *The Guardian* 22 November. www.theguardian.com/world/2017/nov/22/gibraltar-heading-for-abrupt-exit-from-single-market-says-spain (accessed 25 November 2017).

Boffey, D. and O'Carroll, L. (2017) UK plan to register EU citizens would be illegal, say MEPs. *The Guardian* 23 October. www.theguardian.com/politics/2017/oct/23/uk-plan-to-register-eu-citizens-would-be-illegal-say-meps (accessed 23 October 2017).

Bolt, D. (2018) *An Inspection of Exit Checks*. London: Chief Inspector of Borders and Immigration. www.gov.uk/government/news/inspection-report-published-the-home-offices-exit-checks-programme (accessed 2 April 2018).

Bond, A. J., Fundingsland, M. and Tromans, S. (2016) Environmental impact assessment and strategic environmental assessment in the UK after leaving the European Union. *Impact Assessment and Project Appraisal* 34 (3), 271–274.

Bowcott, O. and Borger, J. (2019) UK suffers crushing defeat in UN vote on Chagos Islands. *The Guardian* 22 May. www.theguardian.com/world/2019/may/22/uk-suffers-crushing-defeat-un-vote-chagos-islands (accessed 23 May 2019).

Bowcott, O. and Jones, S. (2015) UN ruling raises hope of return for exiled Chagos islanders. *The Guardian* 19 March. www.theguardian.com/world/2015/mar/19/un-ruling-raises-hope-of-return-for-exiled-chagos-islanders (accessed 20 March 2015).

Box, K. (2018) Nothing green about Brexit's red line. *The New European* 12 April, pp. 8–9.

Boyle, M. and Wood, P. (2017) On maitre d's, Trojan horses and aftershocks: Neoliberalism redux in Ireland after the crash. *Irish Geography* 50 (1), 81–96.

Bozionelos, N., Bozionelos, G., Kostopoulos, K., Shyong, C-H., Baruch, Y. and Zhou, W. (2015) International graduate students' perceptions and interest in international careers. *International Journal of Human Resource Management* 26, 1428–1451.

Brecher, M. (1978) A theoretical approach to international crisis behaviour. *Jerusalem Journal of International Relations* 3 (203), 5–24.

Brinkley, D. (1990) Dean Acheson and the 'Special relationship': The West Point speech of December 1962. *The Historical Journal* 33 (3), 599–608.

Bristow, A., Frenzel, F., Kerr, R. and Robinson, S. (2016) CMS 2017. Brexiting CMS: Critically studying Brexit and its consequences. Liverpool: Edgehill University. www.edgehill.ac.uk/business/files/2016/11/48-brexit-cms-call-final.pdf (accessed 21 November 2017).

British Academy (2017) *Brexit Means …? The British Academy's Priorities for the Humanities and Social Sciences in the Current Negotiations*. London: British Academy. www.britac.ac.uk/ (accessed 25 November 2017).

British Council (2016) *Brexit and Languages: A Checklist for Government Negotiators and Officials*. London: British Council. www.gov.uk/appgmfl-mflbrexit_oct16.pdf (accessed 25 November 2017).

British Council (2017) *Languages for the Future*. London: British Council. www.britishcouncil.org/sites/default/files/languages-for-the-future-report.pdf (accessed 2 March 2018).

British Hospitality Association (2017) *Submission to House of Commons Migration Advisory Committee*. London.

British in Italy (2017) Welcome! About us. www.britishinitaly.net/about-us (accessed 24 February 2018).

Brooks, R. and Waters, J. (2011) *Student Mobilities, Migration and the Internationalization of Higher Education*. Basingstoke: Palgrave Macmillan.

Burns, C., Gravey, V. and Jordan, A. (2018) *UK Environmental Policy Post-Brexit: A Risk Analysis*. London: Friends of the Earth. https://friendsoftheearth.uk/brexit/uk-environmental-policy-post-brexit.pdf (accessed 22 May 2018).

Busby, E. (2017) £10m Mandarin schools programme reaches less than a third of target half way through. *Times Education Supplement* 18 October. www.tes.com/news/ps10m-mandarin-schools-programme-reaches-less-third-target-half-way-through (accessed 8 November 2018).

Butcher, L. (2016) *Regional Airports*. London: House of Commons Briefing Paper SN00323, 26 April.

Butler, S. (2019) Storage costs soar as Brexit stockpiling leads to shortage of warehouse space. *The Guardian* 21 January, p. 31.

Byler, D. (2019) China's hi-tech war on its Muslim minority. *The Guardian* 11 April, Journal pp. 9–11.

Cabinet Office (UK) (2011) Red Tape Challenge. London: HM Government press release, 7 April. www.gov.uk/government/news/red-tape-challenge (accessed 3 March 2018).

Cadwalladr, C. (2018) It's a fight for the soul of our electoral system. *The Observer* 8 July, pp. 34–35.

Cai, L.A., O'Leary, J. and Boger, C. (2000) Chinese travellers to the United States – an emerging market. *Journal of Vacation Marketing* 6 (2), 131–144.

Cain, S. (2018) festival authors 'put off by rise in visa refusals'. *The Guardian* 9 August, pp. 1, 10.

Calder, S. (2017) Brexit travel effects: Everything you wanted to know about holidays after UK leaves the EU. *The Independent* 17 March. www.independent.co.uk/travel/news-and-advice/brexit-travel-effects-holidays-uk-leave-eu-european-union-airline-flights-prices-passports-alcohol-a7633986.html (accessed 8 August 2017).

Camagni, R. and Capello, R. (2015) Rationale and design of EU cohesion policies in a period of crisis. *Regional Science Policy and Practice* 7 (1), 25–49.

Cañas, J.A. (2017) Stoic Gibraltar slowly comes to terms with life after Brexit. *El Pais* 30 March (trans. N. Lyne). http://elpais.com/elpais/2017/03/30/inenglish/1490861341_913797.html (accessed 5 December 2017).

Careers and Enterprise Company (2018) Who we are. London: Careers and Enterprise Company. www.careersandenterprise.co.uk/about-us (accessed 12 November 2018).

Carrington, D. (2018a) European Commission takes UK to court over failure to cut pollution. *The Guardian* 18 May, p. 8.

Carrington, D. (2018b) Tories' track record does not bode well for this dream plan. *The Guardian* 12 January, p. 6.

Carrington, D. (2019) Air pollution falling in London but millions still exposed. *The Guardian* 1 April. www.theguardian.com/environment/2019/apr/01/air-pollution-falling-london-millions-still-exposed (accessed 2 April 2019).

Carroll, R. (2019) While all eyes are on border, Dublin port is preparing for no deal. *The Guardian* 18 March, p. 9.

Carter, H. (2010) Liverpool profited from year as capital of culture, says report. *The Guardian* 12 March, p. 12.

Casey, D. (2019) China-Western Europe aviation market continues to grow. *Routes Online* (Manchester UK) 17 April. www.routesonline.com/news/29/breaking-news/283908/china-western-europe-aviation-market-continues-to-grow/ (accessed 15 June 2019).

Cassidy, K., Innocenti, P. and Bürkner, H-J. (2018) Brexit and new autochthonic politics of belonging. *Space and Polity* 22 (2), 188–204.

Cavlek, N. (2002) Tour operators and destination safety. *Annals of Tourism Research* 29 (2), 478–496.

CBI (2018) *Open and Controlled: a New Approach to Immigration after Brexit*. London: CBI. www.cbi.org.uk/news/open-and-controlled-recommendations-for-a-new-approach-to-immigration/ (accessed 11 August 2018).

CEC (Commission of the European Community) (2007) *Agenda for a Sustainable and Competitive European Tourism*. Brussels: CEC.

Chen, W., Los, B., McCann, P., Ortega-Argilés, R., Thissen, M. and van Oort, F. (2018) The continental divide? Economic exposure to Brexit in regions and countries on both sides of The Channel. *Papers in Regional Science* 97 (1), 25–54.

Chen, Y.Q. (2016) Brexit boost for Chinese students. *China Daily* 30 December. www.chinadaily.com.cn/world/2016-12/30/content_27846271.htm (accessed 10 January 2017).

Chi, R. (2018) Selling the UK university experience to Chinese students. *Emerging Communications*.www.emergingcomms.com/selling-uk-university-experience-chinese-students (accessed 18 March 2018).

Chislett, W. (2017) Why Spain would like a 'soft Brexit' for the UK. *Gibraltar Chronicle* 17 April. http://chronicle.gi/2017/04/why-spain-would-like-a-soft-brexit-for-the-uk/ (accessed 5 December 2017).

Cirer-Costa, J.C. (2017) Turbulence in Mediterranean tourism. *Tourism Management Perspectives* 22, 27–33.

Clegg, P. (2016a) Brexit and the Overseas Territories: repercussions for the periphery. *The Round Table* 105 (5), 543–555.

Clegg, P. (2016b) *The United Kingdom Overseas Territories and the European Union: Benefits and Prospects. Part I – EU benefits to the United Kingdom Overseas Territories*. Bristol, UK: United Kingdom Overseas Territories Association. http://eprints.uwe.ac.uk/29050/ (accessed 16 May 2018).

Clegg, P. (2017) Pitcairn. *The Contemporary Pacific* 29 (1), 166–172.

Cocean, P. and David, N. (2011) Tourism and crisis. *Studia Universitatis Babes-Bolyai, Geographia* 2, 215–222.

Cody, N. (2017) Opening statement. Dublin: Committee on Finance, Public Expenditure and Reform, and Taoiseach, 25 May. www.oireachtas.ie/parliament/media/committees/finance/2017/Niall-Cody,-Chairman,-revenue-Commissioners-Statement.pdf (accessed 4 December 2017).

Coles, T. (2013) Much ado about nothing? Tourism and the financial crisis. In G. Visser and S. Ferreira (eds) *Tourism and Crisis* (pp. 35–52). London: Routledge.

Collins, S.D. (2017) Europe's united future after Brexit: Brexit has not killed the European Union, rather it has eliminated the largest obstacle to EU consolidation. *Global Change, Peace and Security* 29 (3), 311–316.

Collinson, P. and Jones, R. (2016) The post-Brexit pound: how sterling's fall affects you and the UK economy. *The Guardian* 19 August. www.theguardian.com/business/2016/aug/19/the-post-brexit-pound-how-sterlings-fall-affects-the-uk-economy (accessed 20 August 2017).

Commonwealth Secretariat (2015) *The Commonwealth in the Unfolding Global Trade Landscape: Prospects, Priorities, Perspectives*. London: Commonwealth Secretariat.

Commonwealth Secretariat (2016) *Brexit: Its Implications and Potential for the Commonwealth*. London: Commonwealth Secretariat.

Comptroller and Auditor General (2018) *Exiting the EU: the Financial Settlement*. London: National Audit Office.

ComRes (2016) University UK international student poll. Cobham, UK: Communicate Research Ltd. www.comresglobal.com/polls/universities-uk-international-students-poll/ (accessed 18 March 2018).

Conservative Party (2015) *Strong Leadership; A Clear Economic Plan; A Brighter, More Secure Future. The Conservative and Unionist Party Manifesto 2015.* London: Conservative Party.

Conservative Party (2017) *Forward, Together: Our Plan for a Stronger Britain and a Prosperous Future: The Conservative and Unionist Party Manifesto 2017.* London: Conservative Party.

Corbett, A. (2016) Post-Brexit options for UK universities. *University World News* 430. www.universityworldnews.com/article.php?story=20160927183505420 (accessed 18 January 2018).

Cordon, G. and Wardle, S. (2017) Sturgeon brands the return of blue passports 'insular nonsense'. *The Herald* (Glasgow) 23 December, p. 6.

Coughlan, S. (2017) Apprenticeship targets 'poor value for money' says IFS. *BBC News* 31 January. www.bbc.co.uk/news/education-38798305 (accessed 25 November 2017).

Council of the European Communities (1985) Council Directive of 27 June 1985 on the assessment of the effects of certain public and private projects on the environment (85/337/EEC). *Official Journal of the European Communities* C175, 40–49.

Council of the European Union (2019) *The Competitiveness of the Tourism Sector as a Driver for Sustainable Growth, Jobs and Social Cohesion in the EU for the Next Decade.* Brussels: Council of the European Union 9707/19. https://data.consilium. europa.eu/doc/document/ST-9707-2019-INIT/en/pdf (accessed 28 May 2019).

Crabtree, J. (2017) China: Golden week holiday boosts London as tourists cash in on Brexit-hit weak pound. *CNBC* 5 October. www.cnbc.com/2017/10/05/china-golden-week-holiday-boosts-london-as-tourists-cash-in-on-brexit-hit-weak-pound.html (accessed 18 March 2018).

Craggs, R. (2014) Hospitality in geopolitics and the making of Commonwealth international relations. *Geoforum* 52, 90–100.

Craggs, R. (2018) Subaltern geopolitics and the post-colonial Commonwealth, 1965–1990. *Political Geography* 65, 46–56.

Crerar, P. (2018) UK worse off under every Brexit scenario, study finds. *The Guardian* 18 April, p. 13.

Crescenzi, R., Luca, D. and Milio, S. (2016) The geography of the economic crisis in Europe: National macroeconomic conditions, regional structural factors and short-term economic performance. *Cambridge Journal of Regions, Economy and Society* 9 (1), 13–32.

Cró, S. and Martins, A.M. (2017) Structural breaks in international tourism demand: Are they caused by crises or disasters? *Tourism Management* 63, 3–9.

Crotti, R. and Misrahi, T. (2017) *The Travel and Tourism Competitiveness Report 2017.* Geneva: World Economic Forum. www.weforum.org/reports/the-travel-tourism-competitiveness-report-2017 (accessed 22 October 2017).

Curran, R. (2018) Brexit report from Benidorm. *BBC TV Reporting Scotland* 28 March.

Curtice, J. (2017) Scottish public opinion and Brexit: Not so clear after all? In G. Hassan and R. Gunson (eds) *Scotland, the UK and Brexit* (pp. 45–52). Edinburgh: Luath Press.

daa plc (2018) *daa Annual Report 2017.* Dublin: daa.

Dai, B., Jiang, Y., Yang, L. and Ma, Y. (2017) China's outbound tourism – stages, policies and choices. *Tourism Management* 58, 253–258.

Danta, D. and Hall, D. (2000) Introduction. In: D. Hall and D. Danta (eds) *Europe Goes East: EU Enlargement, Diversity and Uncertainty* (pp. 3–14). Norwich: Stationery Office.

Davies, P. (2016) Brexit could cost UK tourism £4.1bn a year, claims study. *Travel Weekly* 1 June. www.travelweekly.co.uk/articles/61811/brexit-could-cost-uk-tourism-41bn-a-year-claims-study (accessed 1 November 2017).

Davis Cross, M.K. and Melissen, J. (2013) *European Public Diplomacy: Soft Power at Work.* Basingstoke, UK: Palgrave Macmillan.

DCMS (UK Department for Digital, Culture, Media and Sport) (2013a) *Business Visits and Events Strategy.* London: DCMS.

DCMS (UK Department for Digital, Culture, Media and Sport) (2013b) *UK campaign to become China's most welcoming tourist destination*. London: DCMS.

DCMS (UK Department for Digital, Culture, Media and Sport) (2016) *Tourism Action Plan*. London: DCMS.

De Falco, F.D. (2019) Cohesion Policy and ESI Funds in 2021–2027: a policy debate. *Eyes on Europe*, 16 June. https://eyes-on-europe.eu/cohesion-policy-and-esi-funds-in-2021-2027-a-policy-debate-cohesion-policy-and-esi-funds-in-2021-2027-a-policy-debate/ (accessed 19 June 2019).

de Vries, C. E. (2017) Benchmarking Brexit: How the British decision to leave shapes EU public opinion. *Journal of Common Market Studies 55*, 38–53.

de Vries, J. (2018) My Amsterdam has been un-created by mass tourism. *The Guardian* 8 August, Journal pp. 10–11.

DEFRA (Department for Environment Food and Rural Affairs) (2016) *Air Pollution in the UK 2015*. London: DEFRA.

DEFRA (Department for Environment Food and Rural Affairs) (2018a) *Agriculture in the UK 2017*. London: DEFRA.

DEFRA (Department for Environment Food and Rural Affairs) (2018b) At a glance: Summary of targets in our 25 year environment plan. London: DEFRA Policy paper.

del Valle Gálvez, A. (2017) Gibraltar and the 'Brexit' – new scenarios within a historic dispute. A proposal. *VerfBlog* 17 April. http://verfassungsblog.de/gibraltar-and-the-brexit-new-scenarios-within-ahistoric-dispute-a-proposal/ (accessed 28 November 2017).

Deliperi, R. (2015) Dean Acherson's observation of Great Britain in 1962. *E-International Relations* 9 August. www.e-ir.info/2015/08/09/ (accessed 24 February 2018).

Deloitte (2016) *What Brexit Might Mean for UK Travel*. London: ABTA.

Dennis, M.J. (2016) Consider implications of Brexit for international student mobility. *Enrollment Management Report* 20 (6), 3.

DftE (Department for the Economy, Northern Ireland (2018) *Background Evidence on the Movement of People Across the Northern Ireland-Ireland Border*. www.economy-ni. gov.uk/sites/default/files/publictions/economy/movement-people-northern-ireland-ireland-border.pdf (accessed 11 May 2018).

DG Internal Policies (2018) *Research for TRAN Committee: BREXIT: Transport and Tourism-The Consequences of a No Deal Scenario*. Brussels: Policy Department for Structural and Cohesion Policies www.europarl.europa.eu/RegData/etudes/STUD/2018/617499/IPOL_STU(2018)617499_EN.pdf (accessed 19 December 2018).

Dick, S. (2019) Scots firms cash in on tourism boom that is full of eastern promise. *The Herald* (Glasgow) 2 February, p. 9.

Dispatcheseuro (2019) No deal, no problem (updated): More EU countries guarantee British expats post-Brexit grace periods. *Dispatcheseurope.com*, 30 May. https://dispatcheseurope.com/no-deal-no-problem-more-eu-countries-guarantee-british-expats-post-brexit-grace-periods/ (accessed 16 June 2019).

Dołżbłasz, S. (2017) From divided to shared spaces: Transborder tourism in the Polish-Czech borderlands. In D. Hall (ed.) *Tourism and Geopolitics: Issues and Concepts from Central and Eastern Europe* (pp. 163–177). Wallingford: CABI.

Dominiczak, P. (2013) Britain must look 'beyond' the EU and focus on links with the Commonwealth. *Daily Telegraph* 25 August. www.telegraph.co.uk/news/politics/10265602/Britain-must-look-beyond-the-EU-and-focus-on-links-with-the-Commonwealth.html (accessed 7 August 2018).

Dowler, L. (2013) Waging hospitality: feminist geopolitics and tourism in West Belfast, Northern Ireland. *Geopolitics* 18 (4), 779–799.

Dredge, D., Gymóthy, S., Birkbak, A., Jensen, T.E. and Madsen, A.K. (2016) *The Impact of Regulatory Approaches Targeting Collaborative Economy in the Tourism Accommodation Sector : Barcelona , Berlin, Amsterdam and Paris*. Copenhagen: Aalborg University, Impulse Paper No 9 for the European Commission DG GROWTH.

Dubow, S. (2017) The Commonwealth and South Africa: From Smuts to Mandela. *Journal of Imperial and Commonwealth History* 45 (2), 284–314.

Dumbrăveanu, D., Tudoricu, A. and Crăciun, A. (2017) European Night of Museums and the geopolitics of events in Romania. In D. Hall (ed.) *Tourism and Geopolitics: Issues and Concepts from Central and Eastern Europe* (pp. 264–279). Wallingford: CABI.

Duncan, P. and O'Carroll, L. (2018) European applications for UK citizenship rise sharply. *The Guardian* 15 March, p. 15.

Durnin, M. (2017) Student mobility: UK markets continued to shrink in 2016. London: British Council. https://siem.britishcouncil.org/insights-blog/student-mobility-uk-markets-continued-shrink-2016 (accessed 18 March 2018).

Dutt, C., Ninov, I. and Haas, D. (2016) The effect of visiting friends and relatives on expatriates' destination knowledge. In V. Katsoni (ed.) *Cultural Tourism in a Digital Age* (pp. 325–327). Berlin: Springer.

Eccles, C. (2011) How middle Britons face downturn. *Destination* 40, 1.

ECFR (European Council on Foreign Relations) (2016) *Brits Abroad: How Brexit Could Hurt Ex-pats.* London: ECFR. www.ecfr.eu/page/-/ecfr_176_-_how_brexit_could_hurt_expats.pdf (accessed 14 May 2018).

ECREU (2018) ECRUE – Expat Citizen Rights in EU. www.ecreu.com (accessed 21 February 2018).

Edwards, C. (2017a) Amatrice mayor tells tourists stop taking selfies in quake rubble. *The Local* (Stockholm) 18 April. www.thelocal.it/20170418/amatrice-mayor-tells-tourists-stop-taking-selfies-in-quake-rubble (accessed 15 May 2018).

Edwards, C. (2017b) Italy least competitive in G7. *The Local* (Stockholm) 28 September. www.thelocal.it/20170928/bureacracy-tax-italy-least-competitive-g7-world-economic-forum (accessed 15 May 2018).

Edwards, E. (2018) Brexit continues to be dominant concern for Irish tourism. *The IrishTimes* 14 August. www.irishtimes.com/news/ireland/brexit-continues-to-be-dominant-concern-for-irish-tourism-1.3596127 (accessed 14 August 2018).

Edwards, R. (2016) Protesters against Brexit gather in Cavan and Fermanagh. *The Irish Times* 8 October. www.irishtimes.com/news/politics/protesters-against-brexit-gather-in-cavan-and-fermanagh-1.2822401 (accessed 6 May 2018).

EERC (2015) About us. www.eerc.org.uk/about-us/ (accessed 21 February 2018).

Electoral Commission (2018) Vote Leave fined and referred to the police for breaking electoral law. London: Electoral Commission, 17 July. www.electoralcommission.org.uk/i-am-a/journalist/electoral-commission-media-centre/party-and-election-finance-to-keep/vote-leave-fined-and-referred-to-the-police-for-breaking-electoral-law (accessed 20 July 2018).

Elliott, L. and Kollewe, J. (2018) Pay rises slow despite fall in jobless and EU worker flight. *The Guardian* 15 August, p. 27.

Emerson, M., Busse, M., Salvo, M. Di, Gros, D. and Pelkmans, J. (2017) *An Assessment of the Economic Impact of Brexit on the EU27.* Brussels: Centre for European Policy Studies, 60pp. www.europarl.europa.eu/RegData/etudes/STUD/2017/595374/IPOL_STU(2017)595374_EN.pdf (accessed 18 March 2018).

Entman, R.M. (2008) Theorizing mediated public diplomacy: the U.S. case. *The International Journal of Press/Politics* 13 (2), 87–102.

Erlanger, S. (2017) No one knows what Britain is anymore. *New York Times*, 4 November, p. SR10. www.nytimes.com/2017/11/04/sunday-review/britain-identity-crisis.html (accessed 24 February 2018).

Estol, J. and Font, X. (2016) European tourism policy: Its evolution and structure. *Tourism Management* 52, 230–241.

European Commission (2010) Europe, the world's No 1 tourist destination – a new political framework for tourism in Europe. *Com(2010) 352*,–14pp.

European Commission (2014) *Results of the Public Consultation "European Tourism of the Future"*. Brussels: European Commission DG Enterprise and Industry, 33pp. https://ec.europa.eu/DocsRoom/documents/7930/attachments/1/translations/en/.../ native (accessed 22 April 2018).

European Commission (2016) *Guide on EU Funding for the Tourism Sector (2014–2020)*. Brussels: European Union, 66pp. https://ec.europa.eu/DocsRoom/documents/18164/ attachments/1/translations/en/.../native (accessed 22 April 2018).

European Commission (2017a) *Joint Report from the Negotiators of the European Union and the United Kingdom Government on Progress During Phase 1 of Negotiations under Article 50 TEU on the United Kingdom's Orderly Withdrawal from the European Union*. Brussels: European Commission.

European Commission (2017b) *White Paper on the Future of Europe: Reflections and Scenarios for the EU27 by 2025*. Brussels: European Commission. https://ec.europa.eu/commission/ sites/.../white_paper_on_the_future_of_europe_en.pdf (accessed 22 April 2018).

European Commission (2018a) *A Stronger, More Efficient and Secure Visa Policy*. Brussels: European Union, Factsheet, 3pp. https://ec.europa.eu/home-affairs/sites/homeaffairs/ files/what-we-do/policies/european-agenda-migration/20180314_eu-visa-policy_ en.pdf (accessed 22 April 2018).

European Commission (2018b) *Commission Staff Working Document Impact Assessment*. Brussels: European Commission, SWD(2018) 77, 90pp. https://ec.europa. eu/.../201780314_ec-staff-working-document-impact-assessment-regulation- establishing-community-code-visas_en.pdf (accessed 22 April 2018).

European Commission (2018c) *EU's Common Visa Policy and Migration*. Brussels: European Commission, News 14 March. https://ec.europa.eu/commission/news/eus-common- visa-policy-and-migration-2018-mar-14_en (accessed 25 March 2018).

European Commission (2018d) *European Strategy for Plastics in a Circular Economy*. Brussels: European Commission. http://eur-lex.europa.eu/legal-content/en/txt/?qid= 1516265440535&uri=com:2018:28:fin (accessed 29 May 2018).

European Commission (2018e) *Internal Preparatory Discussions on Framework for Future Relationship Regulatory Issues*. Brussels: Task Force for the Preparation and Conduct of the Negotiations with the United Kingdom under Article 50 TEU, TF50 (2018) 24, 17 January.

European Commission (2018f) *Overview of EU Tourism Policy*. Brussels: European Commission. https://ec.europa.eu/growth/sectors/tourism/policy-overview_en (accessed 28 March 2018).

European Commission (2018g) *Preparing for the Withdrawal of the United Kingdom from the European Union on 30 March 2019: A Contingency Action Plan*. Brussels: European Commission COM/2018/880final, 13 November.

European Commission (2018h) *Preparing for the Withdrawal of the United Kingdom from the European Union on 30 March 2019: Implementing the Commission's Contingency Action Plan*. Brussels: European Commission COM(2018)890, 19 December.

European Commission (2018i) *Proposal for a Regulation of the European Parliament and the Council amending Regulation (EC) No 810/2009 establishing a Community Code on Visas (Visa Code)*. Brussels: European Commission, SWD(2018) 77 and 78. https:// eur-lex.europa.eu/legal-content/EN/TXT/?uri=COM%3A2018%3A252%3AFIN (accessed 22 April 2018).

European Commission (2018j) *Schengen, Borders and Visas*. Brussels: European Commission, Directorate-General for Migration and Home Affairs.https://ec.europa.eu/home-affairs/ what-we-do/policies/borders-and-visas_en (accessed 24 October 2018).

European Commission (2018k) *Withdrawal of the United Kingdom and EU Rules in the Field of Air Transport*. Brussels: European Commission, Directorate-General for Mobility and Transport, 19 January.

European Commission (2018l) *Withdrawal of the United Kingdom and EU Rules on Consumer Protection and Passenger Rights*. Brussels: European Commission, Directorate-General Justice and Consumers, Directorate-General for Mobility and Transport, 27 February.

European Commission (2018m) *2050 Long-term Strategy*. Brussels: European Commission, 28 November. https://ec.europa.eu/clima/policies/strategies/2050_en (accessed 19 January 2019).

European Commission (2019) *Roadmap to an Agreement on the Union's Long-term Budget for 2011-2027*. Brussels: European Commission COM (2019) 295 final. https://ec.europa.eu/commission/sites/beta-political/files/mff-communication-roadmap-agreement_en.pdf (accessed 19 June 2019).

European Parliament (2017–18) Brexit impact studies. London: European Parliament Liaison Office. www.europarl.europa.eu/unitedkingdom/en/ukevents/brexitstudies.html (accessed 10 April 2018).

European Parliament (2019) *Committee on Transport and Tourism Activity Report 2014–2019*. Brussels: European Parliament. www.europarl.europa.eu/cmsdata/161882/TRAN%20ACTIVITY%20REPORT%202014-2019_19%2003%202019_web.pdf (accessed 28 May 2019).

European Parliament Liaison Office in Ireland (2017) Border Communities Against Brexit awarded European Citizen's Prize. Dublin: European Parliament Liaison Office in Ireland press release 14 July. www.europarl.europa.eu/ireland/en/news-press/border-communities-against-brexit-awarded-european-citizen-s-prize (accessed 6 May 2018).

Evans-Pritchard, A. (2018) Europe's crisis deepens as intellectual opinion turns, and Italy is where it all ends. *The Telegraph Online* 14 March. www.telegraph.co.uk/business/2018/03/14/europes-crisis-deepens-intellectual-opinion-turns-italy-ends/ (accessed 15 May 2018).

Fabbrini, F. (2016) How Brexit opens a window of opportunity for treaty reform in the EU. *Spotlight Europe* 01, 1–8. www.bertelsmann-stiftung.de/de/publikationen/publikation/did/how-brexit-opens-a-window-of-opportunity-for-treaty-reform-in-the-eu/ (accessed 25 March 2018).

Fahy, N., Hervey, T., Greer, S., Jarman, H., Stuckler, D., Galsworthy, M. and McKee, M. (2017) How will Brexit affect health and health services in the UK? Evaluating three possible scenarios. *The Lancet* 390 (10107), 2110–2118.

Fan, Y. (2008) Soft power: Power of attraction or confusion? *Place Branding and Public Diplomacy* 4 (2), 147–158.

Farmaki, A. (2015) Regional network governance and sustainable tourism. *Tourism Geographies* 17 (3), 385–407.

Farrell, H. (2016) British expats in Italy react to Brexit. *The Florentine* (Florence, Italy) 27 June. www.theflorentine.net/lifestyle/2016/06/expats-in-italy-brexit/ (accessed 14 May 2018).

Fearnley-Whittingstall, H. (2018) The plastic crisis? May has left it to Iceland. *The Guardian* 22 January, Journal p. 5.

Fees, C. (1996) Tourism and the politics of authenticity in a North Cotswold town. In: T. Selwyn (ed.) *The Tourist Image: Myth and Myth Making in Tourism* (pp. 121–146). Chichester: John Wiley & Son.

Finch, T. (2010) *Global Brit: Making the Most of the British Diaspora*. London: Institute for Public Policy Research.

Findlay, A., Prazeres, L., McCollum, D. and Packwood, H. (2017) 'It was always the plan': International study as 'learning to migrate'. *Area* 49 (2), 192–199.

Finn, C. (2018) World's largest roll-on/roll-off vessel christened at Dublin Port. *thejournal.ie* 20 April. www.thejournal.ie/ship-launching-celine-dublin-port-3969187-Apr2018/ (accessed 17 November 2018).

Firenzeturismo (2018) Enjoy, respect Firenze: Respect the city. *Firenzeturismo* (Florence, Italy) 6 March. www.firenzeturismo.it/en/news-eventi-2/3284-enjoyrespectfirenze-respect-the-city.html (accessed 15 May 2018).

Ford, M. (2016) The effect of Brexit on workers' rights. *King's Law Journal* 27 (3), 398–415.

Franklin, A. (2004) Tourism as an ordering: Towards a new ontology of tourism. *Tourist Studies* 4 (3), 277–301.

Frost, T. (2017) Book review: 'Brexit: Sociological Responses'. *European Journal of Cultural and Political Sociology* 4 (4), 487–493.

Frost, W. and Hall, C. M. (eds) (2009) *Tourism and National Parks: International Perspectives on Development, Histories, and Change.* Abingdon: Routledge.

Galbraith, J. (2017) Europe and the world after Brexit. *Globalizations* 14 (1), 164–167.

Garcia, J. (2002) The Rock gets rolled. *The National Interest* 69, 119–126.

Gardner, N. and Kries, S. (2017) Funding regional airlines. *Hidden Europe* 7 February. www.hiddeneurope.co.uk/funding-regional-air-services (accessed 5 January 2019).

Garicano, L. (2016) How will Spain negotiate Brexit? Preserving a tangled web. In C. Wyplosz (ed.) *What To Do With the UK? EU Perspectives on Brexit* (pp. 125–133). London: CEPR Press.

Gay, O. and Horton, L. (2011) *Threshold in Referendums.* London: House of Commons Library Standard Note SN/PC/02809.

Gentile, M. and Marcińczak, S. (2014) Housing inequalities in Bucharest: Shallow changes in hesitant transition. *GeoJournal* 79 (4), 449–465.

Geoghegan, P. (2018) Why is Britain turning blind eye to Leave side's lawbreaking? *The Irish Times* 18 July. www.irishtimes.com/opinion/why-is-britain-turning-blind-eye-to-leave-side-s-lawbreaking-1.3568256 (accessed 20 July 2018).

Germain, S. (2018) Health law outside its traditional frontiers: 'trading' medical tourism for just health care in the post-Brexit context. In L. Khoury, C. Regis and R. Kouri (eds) *Health Law at the Frontiers* (in press). Montreal: Yvon Blais, Thomson Reuters.

Getz, D. (2007) *Event Studies: Theory, Research and Policy for Planned Events.* Oxford: Elsevier Butterworth-Heinemann.

Ghosh, H. (2017) *Greatest opportunity in a generation for our countryside to thrive.* London: National Trust.

Gibson, J. (2014) Understanding the tour operators' margin scheme. London: Moore and Smalley. www.mooreandsmalley.co.uk/latest-blogs/understanding-tour-operators-margin-scheme-toms/ (accessed 10 April 2018).

Glasson, J., Therivel, R. and Chadwick, A. (2012) *Introduction to Environmental Impact Assessment.* London: Routledge.

Goodhart, D. (2017) *The Road to Somewhere: The Populist Revolt and the Future of Politics.* London: Hurst.

Goodwin, M. and Milazzo, C. (2017) Taking back control? Investigating the role of immigration in the 2016 vote for Brexit. *British Journal of Politics and International Relations* 19 (3), 450–464.

Gordon, T. (2017) Brexit brinkmanship between Edinburgh and London. *The Herald* (Glasgow) 12 August, p. 15.

Gottlieb, C. (2013) Residential short-term rentals: Should local governments regulate the 'industry'? *Planning and Environmental Law* 65 (2), 4–9.

Government Europa (2019) EU policy recommendations for 2019-24 announced. *Government Europa* 1 May. www.governmenteuropa.eu/eu-policy-recommendations/93112/ (accessed 11 June 2019).

Government of Gibraltar (2016a) *Supplementary Written Evidence (GLT0001).* London: House of Lords European Union Committee. http://data.parliament.uk/writtenevidence/committeeevidence.svc/evidencedocument/european-union-committee/brexit-gibraltar/written/45032.html (accessed 25 October 2018).

Government of Gibraltar (2016b) *Tourist Survey Report 2015.* Gibraltar: Statistics Office.

Government of Gibraltar (2017) *Tourist Survey Report 2016.* Gibraltar: Statistics Office.

Government of Gibraltar (2018a) *Tourist Survey Report 2017.* Gibraltar: Statistics Office.

Government of Gibraltar (2018b) *Tourism 2018*. Gibraltar: Government of Gibraltar.

Government of the Netherlands (2019) Brexit: where do we stand? Government of the Netherlands. www.government.nl/topics/brexit/brexit-where-do-we-stand (accessed 16 June 2019).

Grant, H. (2019) 'Prejudiced' Home Office refuses visas for academics. *The Observer* 9 June, pp. 1–2.

Grant, W. (2016) The challenges facing UK farmers from Brexit. *EuroChoices* 15 (2), 11–16.

Green, D.A (2017) Brexit and Gibraltar. *Financial Times* 6 March. http://blogs.ft.com/ david-allen-green/2017/03/06/brexit-and-gibraltar/ (accessed 5 December 2017).

Green Alliance (2016) The role of European policy in UK environmental quality. Letter to The Right Honourable Liz Truss MP, Secretary of State for Environment, Food and Rural Affairs. www.green-alliance.org.uk/EU_letter_to_Liz_Truss_MP.php (accessed 1 April 2018).

Greener UK (2018) Greener UK. http://greeneruk.org/ (accessed22 May 2018).

Griffith, P. and Morris, M. (2017) *An Immigration Strategy for the UK*. London: Institute for Public Policy Research, Discussion Paper.

Guldi, J. (2017) The case for Utopia: History and the possible meanings of Brexit a hundred years on. *Globalizations* 14 (1), 150–156.

Gustafson, P. (2008) Transnationalism in retirement migration: The case of North European retirees in Spain. *Ethnic and Racial Studies* 31 (3), 451–475.

Haigh, N. (2018) *The Single Market and the Environment. What Kind of Access after Brexit?* Brussels: Institute for European Environmental Policy.

Halkier, H. (2010) EU and tourism development: Bark or bite? *Scandinavian Journal of Hospitality and Tourism* 10 (2), 92–106.

Hall, C.M. (2010) Crisis events in tourism: Subjects of crisis in tourism. *Current Issues in Tourism* 13, 401–417.

Hall, C.M. (2015) Economic greenwash: On the absurdity of tourism and green growth. In V. Reddy and K. Wilkes (eds) *Tourism in the Green Economy* (pp. 339–358). London: Earthscan.

Hall, C.M., James, M. and Baird, T. (2011) Forests and trees as charismatic mega-flora: Implications for heritage tourism and conservation. *Journal of Heritage Tourism* 6 (4), 309–323.

Hall, D.R. (1981) A geographical approach to propaganda. In A.D. Burnett and P.J. Taylor (eds) *Political Studies from Spatial Perspectives* (pp. 313–339). Chichester: John Wiley & Sons.

Hall, D. (2017) Tourism in the geopolitical construction of Central and Eastern Europe (CEE). In: D. Hall (ed.) *Tourism and Geopolitics: Issues and Concepts from Central and Eastern Europe* (pp. 25–35). Wallingford: CABI.

Hall, D. and Brown, F. (2012) The welfare society and tourism: European perspectives. In S. McCabe, L. Minnaert and A. Diekman (eds) *Social Tourism in Europe: Theory and Practice* (pp. 108–121). Bristol: Channel View Publications.

Hall, D., Smith M. and Marciszewska, B. (2006) Introduction. In D. Hall, M. Smith and B. Marciszewska (eds) *Tourism in the New Europe: the Challenges and Opportunities of Enlargement* (pp. 3–17). Wallingford: CABI Publishing.

Hands, S. and Hudson, M. D. (2016) Incorporating climate change mitigation and adaptation into environmental impact assessment: A review of current practice within transport projects in England. *Impact Assessment and Project Appraisal* 34 (4), 330–345.

Hannam, K. (2013) 'Shangri-La' and the new 'Great Game': Exploring tourism geopolitics between China and India. *Tourism, Planning and Development* 10 (2), 178–186.

Hannam, K. (2017) Brexit, Scotland and tourism. In G. Hassan and R. Gunson (eds) *Scotland, the UK and Brexit* (pp. 91–94). Edinburgh: Luath Press.

Hannan, D. (2016) The six best reasons to vote leave. *The Spectator*, 11 June. www. spectator.co.uk/2016/06/six-best-reasons-vote-leave/ (accessed 7 August 2018).

Hansen, P. and Jonsson, S. (2011) Bringing Africa as a 'dowry to Europe'. *Interventions: International Journal of Postcolonial Studies* 13 (3), 443–463.

Hansen, P. and Jonsson, S. (2012) Imperial origins of European integration and the case of Eurafrica: A reply to Gary Marks' 'Europe and its empires'. *Journal of Common Market Studies* 50 (6), 1028–1041.

Hardill, I., Spradbery, J., Arnold-Boakes, J. and Marrugat, M.L. (2005) Severe health and social care issues among British migrants who retire to Spain. *Ageing & Society* 25, 769–783.

Hare, P. (2017) Overseas and overlooked. What now for forgotten corners of British territory? *The New European* 12 October, pp. 23–25.

Harris, J. (2017a) Brexit won't punish bankers. But it will harm voters. *The Guardian* 11 August, p. 29.

Harris, J. (2017b) Whatever happened to the left-behind? *The Guardian* 6 November, p. 27.

Harvey, F. (2019) New UK air pollution curbs 'exceed EU requirements'. *The Guardian* 14 January, p. 2.

Hawkins, O. (2016) *Migration Statistics*. London: House of Commons Library, 6 May. www.parliament.uk/briefing.papers/sn06077.pdf (accessed 5 August 2018).

Haxton, P. (1999) Community participation in the mega-event hosting process: The case of the Olympic Games. Unpublished PhD thesis, University of Technology Sydney (UTS), Australia.

Hazbun, W. (2004) Globalisation, reterritorialisation and the political economy of tourism development in the Middle East. *Geopolitics* 9 (2), 310–341.

Henley, J. (2018) 'It's been just mad'. Property prices boom in Dordogne as Britons dash for the exit. *The Guardian* 28 March, pp. 18–19.

Hereźniak, M. and Florek, M. (2018) Citizen involvement, place branding and mega events: Insights from Expo host cities. *Place Branding and Public Diplomacy* 14 (2), 89–100.

Herrero, L.C., Sanz, J.Á., Devesa, M., Bedate, A. and Barrio, M.J.D. (2006) The economic impact of cultural events: a case-study of Salamanca 2002, European Capital of Culture. *European Urban and Regional Studies* 13, 41–57.

HESA (Higher Education Statistics Agency) (2018a) Figure 7 - HE student enrolments at HE providers in the UK by level of study, mode of study and domicile. London: HESA. www.hesa.ac.uk/data-and-analysis/sfr247/figure-7 (accessed 18 March 2018).

HESA (Higher Education Statistics Agency) (2018b) Top ten European Union countries of domicile (excluding the UK) in 2016/17 for HE student enrolments. London: HESA. www.hesa.ac.uk/data-and-analysis/sfr247/figure-11 (accessed 18 March 2018).

Hewish, T. (2014) *How to Solve a Problem Like a Visa*. London: Commonwealth Exchange. www.commonwealth-exchange.org/how-to-solve-a-problem-like-a-visa/ (accessed 5 August 2018).

Hickey, S. (2019) Brits hedge holiday bets by booking up Bulgaria. *The Observer* 13 January, p. 9.

Highman, L. (2017) *Brexit and the Issues Facing UK Higher Education*. London: UCL Institute of Education, Centre for Global Higher Education, Policy Briefing 2. www.researchcghe.org/publications/Brexit-and-the-issues-facing-UK-higher-education/ (accessed 10 January 2018).

Hirsch, A. (2018) What is the Commonwealth if not the British Empire 2.0? *The Guardian* 18 April, Journal p. 3.

Hobolt, S.B. (2016) The Brexit vote: A divided nation, a divided continent. *Journal of European Public Policy* 23 (9), 1259–1277.

Hobsons (2017) *International Student Survey 2017: Welcoming the World*. London: Hobsons. www.hobsons.com/res/Whitepapers/Hobsons_ISS2017_Global_Welcoming_the_World.pdf (accessed 20 March 2018).

Hogarth, R. and Lloyd, L. (2017) *Who's Afraid of the ECJ? Charting the UK's Relationship with the European Court*. London: Institute for Government.

Home Office (2018a) Apply for a UK residence card. www.gov.uk/apply-for-a-residence-card (accessed 21 February 2018).

Home Office (2018b) UK residence for EU citizens. www.gov.uk/eea-registration-certificate/permanent-residence (accessed 21 February 2018).

Hopkins, N. (2019) Revealed: secret plan for months of no deal chaos. *The Guardian* 23 March, pp. 1–2.

House of Commons (2018) *Exiting the European Union Committee. EU Exit Analysis Cross Whitehall Briefing*. London: House of Commons. www.parliament.uk/... committees/Exiting-the-European-Union/EU-Exit-Analysis-Cross-Whitehall-Briefing. pdf (accessed 24 April 2018).

House of Commons Library (2018) *Brexit and the Environment*. Briefing Paper CBP8132, 19 January. London: House of Commons Library.

House of Lords (2017) *European Union Committee 13th. Report of Session 2016–17: Brexit – Gibraltar*. HL Paper 116, 1 March. London: House of Lords.

Howell, D. (2016) Brexit, the UK and the Commonwealth: opportunities and challenges. *The Round Table* 105 (5), 575–576.

Huang, R. (2005) The experience of Mainland Chinese international students in the UK. Unpublished PhD thesis. Derby, UK: University of Derby.

Huang, R. (2008) Mapping educational tourists experience in the UK. *Third World Quarterly* 29 (5), 1003–1020.

Huang, R. (2013) International experience and graduate employability: Perceptions of Chinese international students in the UK. *Journal of Hospitality, Leisure, Sport & Tourism Education* 13, 87–96.

Huang, R. and Tian, X.R. (2013) An investigation of travel behaviour of Chinese international students in the UK. *Journal of China Tourism Research* 9 (3), 277–291.

Huang, R. and Turner, R. (2018) International experience, universities support and graduate employability: Perceptions of Chinese international students studying in UK universities. *Journal of Education and Work* 31 (2), 175–189.

Huang, R., Turner, R. and Chen, Q. (2014) Chinese international students' perspective and strategies in preparing for their future employability. *Journal of Vocational Education and Training* 66 (2), 175–193.

Huang, W-J., Ramshaw, G. and Norman, W. C. (2016) Homecoming or tourism? Diaspora tourism experience of second-generation immigrants. *Tourism Geographies* 18 (1), 59–79.

Hubble, S. and Bolton, P. (2018) *International and EU Students in Higher Education in the UK*. London: House of Commons Briefing Paper CBP 7976.

Hughes, B. (2017) Poll suggests unionists would be content with sea border after Brexit. *The Irish News* (Belfast) 28 November. www.irishnews.com/news/brexit/2017/11/28/news/ poll-suggests-unionists-would-be-content-with-irish-sea-border-after-brexit-1197975/ (accessed 3 April 2018).

Hutton, B. (2018) Ireland has 208 border crossings, officials from North and South agree. The Irish Times 26 April. www.irishtimes.com/news/ireland/irish-news/ireland-has-208-border-crossings-officials-from-north-and-south-agree-1.3474246 (accessed 11 May 2018).

Hyde, M. (2017) It's deal or no deal, Tory style. The prize is a cliff-edge Brexit. *The Guardian* 14 October, p. 34.

Iammarino, S., Rodríguez-Pose, A. and Storper, M. (2017) *Why Regional Development Matters for Europe's Future*. Brussels: European Commission DG Regio, Working Paper 7/2017.

Iannelli, C. and Huang, J. (2014) Trends in participation and attainment of Chinese students in UK higher education. *Studies in Higher Education* 39 (5), 805–822.

ICCA (International Conference and Convention Association) (2017) *2016 ICCA Statistics Report: Country and City Rankings*. Amsterdam: ICCA.

INE (Instituto Nacional de Estadística) (2018) *Movimentos Turísticas en Fronteras: Resultos Nacionales*. Madrid: INE. www.ine.es/jaxiT3/Datos.htm?t=23984 (accessed 11 February 2018).

Inman, P. (2017) 'Fall in confidence' over Brexit as foreign buyers shun UK buyouts. *The Guardian* 9 October, p. 7.

Insch, A. and Avraham, E. (2014) Managing the reputation of places in crisis. *Place Branding and Public Diplomacy* 10 (3), 171–173.

Insights (2018) Chinese tourist market to grow even bigger. eHotelier.com 2 February. https://insights.ehotelier.com/insights/2018/02/02/chinese-tourist-market-grow-even-bigger/ (accessed 13 November 2018).

InterNations (2017) Settling down in La Dolce Vita, Italy. *InterNations* (Munich and Vilnius). www.internations.org/expat-insider/2017/italy-39218 (accessed 15 May 2018).

IPPR (Institute for Public Policy Research) (2018) No public appetite for deregulation post-Brexit according to new polling for IPPR. London: IPPR, Press release 19 February. www.ippr.org/news-and-media/press-releases/no-public-appetite-for-deregulation-post-brexit-according-to-new-polling-for-ippr (accessed 22 May 2018).

Irwin, G. (2015) *BREXIT: the Impact on the UK and the EU*. London: Global Counsel. www.global-counsel.co.uk/sites/.../Global%20Counsel_Impact_of_Brexit.pdf (accessed 22 April 2018).

Jablonowski, K., Olivas Osuna, J.J., De Lyon, J. *et al.* (2018) *Understanding Brexit: Impacts at a Local Level. Ceredigion Case Study*. London: LSE Conflict and Civil Society Research Unit.

Jacobs, M. (2018) Act now to stop Brexit harming the planet. *The Guardian* 4 April, Journal p. 5.

Jacobs, M. and Mazzucato, M. (2016) *Rethinking Capitalism: Economics and Policy for Sustainable and Inclusive Growth*. Chichester: John Wiley & Sons.

Jago, L., Chalip, L., Brown, G., Mules, T. and Shameem, A. (2003) Building events into destination branding: Insights from experts. *Event Management* 8 (1), 3–14.

Janson, K. (2016a) UK prioritises tourism. Presentation to members. London: Tourism Alliance.

Janson, K. (2016b) *Referendum Impact Survey*. London: Tourism Alliance.

Janson, K. (2017) Four policies for post-Brexit tourism. *Tourism* 167, 4–5.

Janta, H. and Christou, A. (2019) Hosting as social practice: Gendered insights into contemporary tourism mobilities. *Annals of Tourism Research* 74, 167–176.

Janta, H. and Ladkin, A. (2013) In search of employment: Online technologies and Polish migrants. *New Technology, Work and Employment* 28 (3), 241–253.

Jary, M. (2017) How the world sees us. *The New European* 23 November, 27–30.

JCQ (Joint Council for Qualifications) (2018) *GCE A & AS Levels: Summer 2018 Results*. London: JCQ.

Jessop, B. (2017) The organic crisis of the British state: Putting Brexit in its place. *Globalizations* 14(1), 133–141.

Johnson, S., Rayner, G. and Hope, C. (2018) Fishermen 'betrayed' by May's Brexit deal. *The Daily Telegraph* 20 March, pp. 1/4.

Jones, S. (2018a) High town: In the shadow of Gibraltar and drug gangs. *The Guardian* 5 April, p. 31.

Jones, S. (2018b) 'We didn't ask for this, but …' Gibraltar faces up to Brexit. *The Guardian* 6 April, p. 24.

Kaneva, N. and Popescu, D. (2011) National identity lite: Nation branding in post-communist Romania and Bulgaria. *International Journal of Cultural Studies* 14 (2), 191–207.

Kantar TNS (2018a) *The GB Tourist 2017 Annual Report*. London: VisitScotland/Visit Wales/VisitEngland. www.visitbritain.org/gb-tourism-survey-2017-overview (accessed 9 October 2018).

Kantar TNS (2018b) *The Great Britain Day Visitor 2017 Annual Report*. London: VisitScotland/Visit Wales/VisitEngland. www.visitbritain.org/gb-day-visits-survey-latest-results (accessed 14 October 2018).

Kastenholz, E. (2010) 'Cultural proximity' as a determinant of destination image. *Journal of Vacation Marketing* 16 (4), 313–322.

Kelly, J. (2016) Brexit: How much of a generation gap is there? London: BBC. www.bbc.co.uk/news/magazine-36619342 (accessed 22 February 2017).

Kent, M-C. (2017) *What to Make of the New Industrial Strategy?* London: People 1st. www.people1st.co.uk/external-blogs/what-to-make-of-the-new-industrial-strategy/ (accessed 28 November 2017).

Kentish, B. (2018) Government's anti-homelessness taskforce has not held a single meeting since it was set up, ministers admit. *The Independent* 20 February. www.independent. co.uk/news/uk/politics/government-homelessness-rough-sleeping-tories-taskforce-not-held-meeting-john-healey-philip-hammond-a8220101.html (accessed 28 February 2018).

Kiambi, D (2017) The role of familiarity in shaping country reputation. In J. Fullerton and A. Kendrick (eds) *Shaping International Public Opinion: a Model for Nation Branding and Public Diplomacy* (pp. 57–75). New York, NY: Peter Lang.

Kiambi, D. and Shafer, A. (2018) Country reputation management: Developing a scale for measuring the reputation of four African countries in the United States. *Place Branding and Public Diplomacy* 14 (3), 175–186.

Kiefel, M., Olivas Osuna, J.J., De Lyon, J. *et al.*(2018) *Understanding Brexit: Impacts at a Local Level. Pendle Case Study.* London: LSE Conflict and Civil Society Research Unit.

Kim, S.E. and Lehto, X.Y. (2012) The voice of tourists with mobility disabilities: Insights from online customer complaint websites. *International Journal of Contemporary Hospitality Management* 24 (3), 451–476.

Kirkup, J. and Winnett, R. (2012) Theresa May interview: 'We're going to give illegal immigrants a really hostile reception'. *The Daily Telegraph* 25 May. www.telegraph. co.uk/news/uknews/immigration/9291483/Theresa-May-interview-Were-going-to-give-illegal-migrants-a-really-hostile-reception.html (accessed 25 April 2018).

Kock, F., Josiassen, A. and Assaf, A.G. (2019) The xenophobic tourist. *Annals of Tourism Research* 74, 155–166.

Kotler, P., Haider, D.H. and Rein, I.J. (1993) *Marketing Places: Attracting Investment, Industry, and Tourism to Cities, States, and Nations.* New York, NY: Free Press.

KPMG (2017) *Brexit: The Impact on Sectors.* London: KPMG Economic Insights, 24 February.

KPMG/BHA (British Hospitality Association) (2017) *Labour Migration in the Hospitality Sector.* London: KPMG Economics.

Kramme, M. (2017) *Consequences of Brexit in the Area of Consumer Protection.* Brussels: European Parliament Committee on the Internal Market and Consumer Protection.

Kroll, D.A. and Leuffen, D. (2016) Ties that bind can also strangle: The Brexit threat and the hardships of reforming the EU. *Journal of European Public Policy* 23 (9), 1311–1320.

Kuo, L. (2018) China insists 1.1m Muslims in internment camps are being educated. *The Guardian* 15 September, p. 39.

Kwek, A., Wang, Y. and Weaver, D.B. (2014) Retail tours in China for overseas Chinese: Soft power or hard sell? *Annals of Tourism Research* 44 (1), 36–52.

Laine, J. (2017) Finnish-Russian border mobility and tourism: Localism overruled by geopolitics. In D. Hall (ed.) *Tourism and Geopolitics: Issues and Concepts from Central and Eastern Europe* (pp. 178–190). Wallingford: CABI.

Lander, E. and Tims, A. (2018) All abroad: Meet the Brexiles who voted with their feet. *The Guardian* 24 March, pp. 51–53.

Lang, T., Lewis, T., Marsden, T. and Millstone, E. (2018) *Feeding Britain: Food Security after Brexit.* London: City University Centre for Food Policy.

Lanouar, C. and Goaied, M. (2019) Tourism, terrorism and political violence in Tunisia: Evidence from Markov-switching models. *Tourism Management* 70, 404–418.

Lazzeretti, L. (2003)City of art as a high local culture system and cultural districtualization processes: The cluster of art restoration in Florence. *International Journal of Urban and Regional Research* 27 (3), 635–648.

Leave.EU (2016) Trade. Bristol: Leave.EU. http://leave.eu/en/trade (accessed 7 August 2018).

Lee, N., Morris, K. and Kemeny, T. (2017) *Immobility and the Brexit vote.* London: LSE International Inequalities Institute Working paper 19.

Lim, W.L. (2017) Restoring tourist confidence and travel intentions after disasters: Some insights from a rejoinder to a series of unfortunate events in Malaysian tourism. *Current Issues in Tourism* 20 (1), 38–42.

Lim, W. M. (2018) Exiting supranational unions and the corresponding impact on tourism: Some insights from a rejoinder to Brexit. *Current Issues in Tourism* 21 (9), 970–974.

Lloyd, M. (2016) EU referendum: Leave's nationalism takes different forms on the right and the left. London: LSE blog 7 April. http://blogs.lse.ac.uk/europpblog/2016/04/07/eu-referendum-leaves-nationalism-takes-different-forms-on-the right-and-on-the-left/ (accessed 5 August 2018).

Local Government Association (2016) *Museum Survey of Heads of Cultural Services.* London: Local Government Association. www.local.gov.uk/museum-survey-heads-cultural-services (accessed 16 August 2018).

Local Government Association (2017a) *Beyond Brexit: Future of Funding Currently Sourced from the EU.* London: Local Government Association.

Local Government Association (2017b) *Employment and Skills: Work Local.* London: Local Government Association. www.local.gov.uk/topics/employment-ans-skills/work-local#2 (accessed 16 August 2018).

Local Government Association (2018) The impact of Brexit on tourism and creative industries. Briefing. London: House of Commons, 17 April. www.LGABriefing-BrexitandTourism_HofC_April18_v2.pdf (accessed 14 August 2018).

Lockyer, T.I.M. and Ryan, C. (2007) Visiting friends and relatives: Distinguishing between the two groups: The case of Hamilton, New Zealand. *Tourism Recreation Research* 32 (1), 59–68.

Loh, S. (2016) As many Britons mourn Brexit, Asian students celebrate weaker pound. *Asian Correspondent*, July. https://asiancorrespondent.com/2016/07/many-britons-mourn-brexit-international-students-celebrate-weaker-pound/#2VCmIRDxMRhdze3z.99 (accessed 10 March 2018).

Los, B., McCann, P., Springford, J. and Thissen, M. (2017) The mismatch between local voting and the local economic consequences of Brexit. *Regional Studies* 51 (5), 786–799.

Lowe, R. (2016) International higher education in facts and figures. London: Universities UK. www.universitiesuk.ac.uk/blog/Pages/international-higher-education-in-facts-and-figures.aspx (accessed 16 January 2018).

Lowenthal, D. (1994) European and English landscapes as national symbols. In D. Hooson (ed.) *Geography and National Identity* (pp.15–38). Oxford: Blackwell.

Luhmann, N. (1982) *The Differentiation of Society.* New York, NY: Columbia University Press.

MacKenzie, J.M. (2016) Brexit: The view from Scotland. *The Round Table* 105 (5), 577–579.

Macleod, D.V.L. and Carrier, J.G. (2010) *Tourism, Power and Culture: Anthropological Insights.* Bristol: Channel View Publications.

Macquisten, G. (2017) *The Shape of Gibraltar in the Aftermath of Brexit.* London: The Bruges Group. www.brugesgroup.com (accessed 28 November 2017).

Maer, L. and Ryan-White, G. (2018) *Exiting the EU: Sectoral Assessments.* London: House of Commons Library, Briefing Paper 08128, 9 March.

Mance, H. (2016) Britain has had enough of experts, says Gove. *The Financial Times* 3 June. www.ft.com/content/3be49734-29cb-11e6-83e4-abc22d5d108c (accessed 17 October 2017).

Manente, M., Minghetti, V. and Montaguti, F. (2013) The role of the EU in defining tourism policies for a competitive destination governance. In C. Costa, E. Panyik and D. Buhalis (eds) *Trends in European Tourism Planning and Organisation* (pp. 208–219). Bristol: Channel View Publications.

Mann, M. (2019) Using the big freeze to deny climate change … stupidity or cynicism? *The Observer* 3 February, p. 41.

Mann, M.E. and Toles, T. (2016) *The Madhouse Effect: How Climate Change Denial Is Threatening Our Planet, Destroying Our Politics, and Driving Us Crazy.* New York, NY and Chichester, UK: Columbia University Press.

Marcińczak, S., Gentile, M., Rufat, S. and Chelcea, L. (2014) Urban geographies of hesitant transition: Tracing socioeconomic segregation in post-Ceaușescu Bucharest. *International Journal of Urban and Regional Research* 38 (4), 1399–1417.

Mardell, J. (2017) Post-Brexit Britain: Educating China. *The China Road* 10 November. http://thechinaroad.co.uk/2017/11/10/post-brexit-britain-educating-china/ (accessed 10 January 2018).

Marr, D. (2017) Brexit will further diminish Britain's global influence. *The Herald* (Glasgow) 9 December, p. 13.

Marschall, S. (2015) 'Homesick tourism': Memory, identity and (be) longing. *Current Issues in Tourism* 18 (9), 876–892.

Marschall, (2017) Transnational migrant home visits as identity practice: The case of African migrants in South Africa. *Annals of Tourism Research* 63, 140–150.

Marshall, P. (2016) Brexit in its worldwide aspect: An opportunity to be grasped. *The Round Table* 105 (5), 451–461.

Martill, B. *(2017) Britain has lost a role, and failed to find an empire. London: UCL European Institute, 17 January.* www.ucl.ac.uk/european-institute/analysis/2016-17/martill-may-speech (accessed 24 February 2018).

Martin, R., Pike, A., Tyler, P. and Gardiner, B. (2016) Spatially rebalancing the UK economy: Towards a new policy model? *Regional Studies* 50 (2), 342–357.

Mason, R. and Walker, P. (2017) EU rules British cities cannot be capitals of culture. *The Guardian* 23 November. www.theguardian.com/politics/2017/nov/23/eu-rules-british-cities-cannot-be-capitals-of-culture (accessed 25 November 2017).

Matos, P. (2006) Hosting mega sports events – a brief assessment of their multidimensional impacts. Paper presented at conference: Economic and Social Impact of Hosting Mega Sports Events, Copenhagen , 1 September.

Matti, J. and Zhou, Y. (2017) The political economy of Brexit: Explaining the vote. *Applied Economics Letters* 24 (16), 1131–1134.

May, A. (2013) The Commonwealth and Britain's turn to Europe. *The Round Table* 102 (1), 29–39.

Mayall, J. (2016) Some reflections on Brexit and its aftermath. *The Round Table* 105 (5), 573–574.

McBride, I. (2016) After Brexit the border will again dominate Ireland. *The Guardian* 20 July, p. 28.

McCann, G. and Hainsworth, P. (2017) Brexit and Northern Ireland: The 2016 referendum on the United Kingdom's membership of the European Union. *Irish Political Studies* 32 (2), 327–342.

McGuinness, T. and Hawkins, O. (2017) *Brexit: What Impact on Those Currently Exercising Free Movement Rights?* London: House of Commons Library Briefing Paper 7871.

McIntyre, W.D. (2016) Brexit: A view from New Zealand. *The Round Table* 105 (5), 591–592.

McKay, S. (2018) How old ghosts are haunting Ireland. *The Observer* 25 March, pp. 53–55.

McKee, D. and McKee, M. (2018) What might Brexit mean for British tourists travelling to the rest of Europe? *Journal of the Royal College of Physicians of Edinburgh* 48 (2), 134–140.

McKinnon, A. and Fransoo, J. (2019) Brexit stockpiles and the bullwhip effect. *The Guardian* 5 January, Journal p. 6.

McSmith, A. (2016) Gibraltar faces 'existential threat' to its economy if there's a 'hard Brexit' deal, its chief minister warns. The Independent 20 August. www.independent.co.uk/news/uk/gibraltar-existential-threat-economy-hard-brexit-deal-eu-fabian-picardo-a7201211.html (accessed 24 October 2018).

Meleady, R., Seger, C.R. and Vermue, M. (2017) Examining the role of positive and negative intergroup contact and anti-immigrant prejudice in Brexit. *British Journal of Social Psychology* 56 (4), 799–808.

Michael, I., Armstrong, A. and King, B. (2003) The travel behaviour of international students: The relationship between studying abroad and their choice of tourist destinations. *Journal of Vacation Marketing* 10 (1), 57–66.

Migration Advisory Committee (2018) EEA Migration in the UK: Final Report. London: Migration Advisory Committee.

Migration Watch (2016) *The British in Europe – and Vice Versa*. London: Migration Watch, MW354, 23 March.

Miller, B. (2018) What proportion of UK law is made in Brussels? London: Kings College London. http://ukandeu.ac.uk/fact-figures/what-proportion-of-uk-law-is-made-in-brussels/ (accessed 24 September 2018).

Miller, V. (2014) *Making EU Law into UK Law*. London: House of Commons, Note SN/1A/7002, 22 October.

Miller, V. (ed.) (2016) *Brexit: Impact Across Policy Areas*. London: House of Commons, Briefing Paper 07213, 26 August.

Mindus, P. (2017) *European Citizenship After Brexit: Freedom of Movement and Rights of Residence*. Basingtoke, UK: Palgrave Macmillan.

Minnaert, L. (2012) The value of social tourism for disadvantaged families. In H. Schänzel, I. Yeoman and E. Backer (eds) *Family Tourism: Multidisciplinary Perspectives* (pp. 93–104). Bristol: Channel View Publications.

Mitchell, T., Zaman, S. and Raja, C. (2018) *Estimating the Potential Impact of Brexit on Commonwealth Tourism, Remittances and Aid*. London: Commonwealth Secretariat, Economics Technical Working Paper 2018/01.

Mody, M., Suess, C. and Dogru, T. (2019) Not in my back yard? Is the anti-Airbnb discourse truly warranted? *Annals of Tourism Research* 74, 198–203.

Mok, K. H., Han, X., Jiang, J. and Zhang, X. (2017) *International and Transnational Learning in Higher Education: a Study of Students' Career Development in China*. London: UCL Institute of Education, Centre for Global Higher Education, Working Paper 21.

Monbiot, G. (2016a) *How Did We Get Into This Mess?* London: Verso.

Monbiot, G. (2016b) The Tories will grind our environment into the dust. *The Guardian* 20 July, p. 29.

Monbiot, G. (2017) Farmers fear life outside the EU, but it could mean a rebirth for rural Britain. *The Guardian* 11 January, p. 31.

Monbiot, G. (2018) May's plastic plan is big on gimmicks but it won't cut waste. *The Guardian* 12 January, p. 34.

Morgan, N., Pritchard, A. and Sedgley, D. (2015) Social tourism and well-being in later life. *Annals of Tourism Research* 52, 1–15.

Morris, C. (2018) Brexit: why services matter in any deal. London: BBC 5 July. www.bbc.co.uk/news/uk-44724376 (accessed 14 July 2018).

Morris, S. and Carell, S. (2018) UK government accused of Brexit power grab by Scots and Welsh. *The Guardian* 27 February, p. 10.

Morrison, N (2019) Brexit uncertainty proving no bar to international students. *Forbes.com* 7 February. www.forbes.com/sites/nickmorrison/2019/02/07/brexit-uncertainty-proving-no-bar-to-international-students/#70b6c2d166c0 (accessed 20 June 2019).

Moscato, D. (2018) Cultural congruency in mediated gastrodiplomacy: A qualitative framing analysis of the U.S.-Japan Sushi Summit. *Place Branding and Public Diplomacy* 14 (3), 187–196.

Moss, S. (2019) 'Britain no longer feels like home' *The Guardian* 5 February, G2 pp. 8–9.

Mueller, D. (2015) Young Germans in England visiting Germany: Translocal subjectivities and ambivalent views of 'home'. *Population, Space and Place* 21 (7), 625–639.

Murphy, A. (2016a) *Brexit and the Irish Tourist Sector*. Dublin: Crowe Howarth.

Murphy, A. (2016b) *Impact of Brexit on the Irish Hotel Sector*. Dublin: Crowe Howarth.

Murphy, A. (2018) *Brexit and the Irish Tourist Sector – March 2018 Update*. Dublin: Crowe Howarth.

Murphy, P. (2013) Time for Britain to rethink its place in the Commonwealth. *The Conversation* 15 November. http://theconversation.com/time-for-britain-to-rethink-its-place-in-the-commonwealth-20327 (accessed 5 August 2018).

Murphy, P. (2018) *The Empire's New Clothes: the Myth of the Commonwealth*. London: Hurst.

Murray, G. (2019) Is the sun starting to set on our love affair with Benidorm? *The Herald* (Glasgow) 5 January, p. 5.

Murray-Evans, P. (2016) Myths of Commonwealth betrayal: UK-Africa trade before and after Brexit. *The Round Table* 105 (5), 489–498.

Mut Bosque, M. (2017) Brexit and the Commonwealth: New challenges for Gibraltar. *The Round Table: The Commonwealth Journal of International Affairs* 106 (4), 483–485.

Namusoke, E. (2016) A divided family: Race, the Commonwealth and Brexit. *The Round Table* 105 (5), 463–476.

Nash, C., Dennis, L. and Graham, B. (2010) Putting the border in place: Customs regulation in the making of the Irish border, 1921–1945. *Journal of Historical Geography* 36, 421–431.

National Trust (2018) Call for Government to support nature friendly farming policies post-Brexit. London: National Trust, Press release, 5 January.

Naughton, J. (2018) The new China confounds everything western liberals believed about the net. *The Observer* 11 November, p. 25.

Nelson, F. (2011) The Euro masquerade. *The Spectator* 26 October. https://blogs.spectator.co.uk/2011/10/the-euro-masquerade/ (accessed 7 August 2018).

Nesbit, M. and Baldock, D. (2018) *Brexit and the Level Playing Field: Key Issues for Environmental Equivalence*. London: Institute for European Environmental Policy.

Neslen, A. (2018) EU tells Britain: Show plan on air quality or go to court. *The Guardian* 31 January, p. 4.

Netto, G. and Craig, G. (2016) Migration and differential labour market participation: Theoretical directions, recurring themes, implications of Brexit and areas for future research. *Social Policy and Society* 16 (4), 613–622.

Neumayer, E. (2006) Unequal access to foreign spaces: How states use visa restrictions to regulate mobility in a globalized world. *Transactions of the Institute of British Geographers* 31 (1), 72–84.

Newry, Mourne and Down District Council (2017) *Newry City Centre Map*. Newry, UK: Newry, Mourne and Down District Council.

NFU (2017) *NFU Labour Force Surveys*. London: NFU.

Nicholson, R. and Marquis, C. (2015) Introduction. In R. Nicholson, C. Marquis and G. Szamosi (eds) *Contested Identities: Literary Negotiations in Time and Place* (pp. xi–xxv). Newcastle-upon-Tyne: Cambridge Scholars Publishing.

North South Ministerial Council (2018) North South Ministerial Council. www.northsouthministerialcouncil.org (accessed 6 May 2018).

Nurković, R. and Hall, D. (2017) Rural tourism as a meeting ground in Bosnia and Herzegovina? In D. Hall (ed.) *Tourism and Geopolitics: Issues and Concepts from Central and Eastern Europe* (pp. 236–249). Wallingford: CABI.

Nye, J.S. (2008) Public diplomacy and soft power. *Annals of the American Academy of Political and Social Science* 616 (1), 94–109.

O'Carroll, L. (2017a) Briton in Spain fears losing access to healthcare after Brexit. *The Guardian* 22 July. www.theguardian.com/politics/2017/jul/22/british-expat-in-spain-fears-losing-access-to-healthcare-after-brexit (accessed 23 October 2017).

O'Carroll, L. (2017b) Ireland cross-border activities 'are at risk'. *The Guardian* 28 November, p. 7.

O'Carroll, L. (2018a) Residency: UK nationals in the dark on their status. *The Guardian* 28 March, p. 19.

O'Carroll, L. (2018b) Ski businesses warn thousands of jobs for Britons are at risk. *The Guardian* 19 December, p. 11.

O'Rourke, K. (2019) *A Short History of Brexit: From Brentry to Backstop.* London: Pelican.

O'Toole, F. (2017) In the new Trump-Brexit global order, Ireland could be ripped apart. *The Guardian* 23 February, p. 31.

O'Toole, F. (2018) The kind of unity Ireland needs isn't about territory – it is about people. *The Observer* 16 September, p. 35.

OECD (Organisation for Economic Cooperation and Development) (2013) *Education Indicators in Focus.* Paris: OECD.

OECD (Organization for European Cooperation and Development) (2016) *Tourism Trends and Policies.* Paris: OECD.

Offord, P. (2018) Careers and Enterprise Company slammed for multimillion-pound research spending. *FEWeek* 16 May. https://feweek.co.uk/2018/05/16/careers-and-enterprise-company-grilled-on-multi-million-pound-research-spending/ (accessed 12 November 2018).

OJEU (Official Journal of the European Union) (2011) *Directive 2011/24/EU of the European Parliament and of the Council 9 March 2011.* Brussels: European Union.

Oliver, C. (2017) *Unleashing Demons.* London: Hodder.

Oliver, T. (2015) How the EU responds to a British withdrawal will be determined by five key factors. *LSE Blogs* (London) 3 December. http://blogs.lse.ac.uk/brexit/2015/12/03/europes-potential-responses-to-a-british-withdrawal-from-the-union-will-be-determined-by-ideas-interests-institutions-the-international-and-individuals/ (accessed 21 November 2017).

Oliver, T. (2016) What impact would a Brexit have on the EU?' *The Dahrendorf Forum* 29(5), 16–20.

Oliver, T. and Boyle, A. (2017) Brexit is a fascinating case study for the next generation of students and teachers of British and European politics. *LSE Blogs* (London) 7 July. http://blogs.lse.ac.uk/brexit/2017/07/07/brexit-teachers-and-brexit-students-a-fascinating-case-study-for-the-social-sciences/(accessed 21 November 2017).

ONS (UK Office for National Statistics) (2017) *What information is there on British migrants living in Europe.* London: ONS.

ONS (UK Office for National Statistics) (2018) *UK Labour Market: November 2018.* London: ONS.

ONS (UK Office for National Statistics) (2019a) *Labour Market Overview UK: May 2019.* London: ONS.

ONS (UK Office for National Statistics) (2019b) *Leisure and Tourism Statistics.* London: ONS.

ONS (UK Office for National Statistics) (2019c) *Overseas Travel and Tourism: February 2019 Provisional Results.* London: ONS.

Onslow, S. (2015) The Commonwealth and the cold war, neutralism and non-alignment. *The International History Review* 37 (5), 1059–1082.

ORC (Office of the Revenue Commissioners) (2016) *Brexit and the Consequences for Irish Customs.* Draft document. Dublin: ORC.

Orwell, G. (1941/1954) The lion and the unicorn. In G. Orwell *England, Your England, and Other Essays.* London: Secker and Warburg, pp. 193–194.

Ostrowski, M.S. (2019) The Rock backstop. *The New European* 7 March, pp. 26–27.

Ousby, I. (1990) *The Englishman's England: Taste, Travel and the Rise of Tourism.* Cambridge: Cambridge University Press.

Packwood, H., Findlay, A. and McCollum, D. (2015) *International Study for an International Career: A Survey of the Motivations and Aspirations of International Students in the UK.* Southampton, UK:ESRC Centre for Population Change, CPC Briefing Paper 27.

Pain, R. and Staeheli, L. (2014) Introduction: Intimacy-geopolitics and violence. *Area* 46 (4), 344–347.

Panyik, E. and Anastasiadou, C. (2013) Mapping the EU's evolving role in tourism: Implications for the new EU competence. In C. Costa, E. Panyik and D. Buhalis (eds) *Trends in European Tourism Planning and Organisation* (pp. 189–207). Bristol: Channel View Publications.

Papatheodorou, A., Rosselló, J. and Xiao, H. (2010) Global economic crisis and tourism: consequences and perspectives. *Journal of Travel Research* 49, 39–45.

Pappas, N. (2019) UK outbound travel and Brexit complexity. *Tourism Management* 72, 12–22.

Patomäki, H. (2017) Will the EU disintegrate? What does the likely possibility of disintegration tell about the future of the world? *Globalizations* 14 (1), 168–177.

Pearce, P.L. (2012) The experience of visiting home and familiar places. *Annals of Tourism Research* 39 (2), 1024–1047.

Pegg, D. and Grierson, J. (2018) Home Office fails to suspend 'golden visa' scheme. *The Guardian* 12 December, pp. 16–17.

Pendleton, A., Salveson, P. and Kiberd, E. (2019) *A Rail Network for Everyone: Probing HS2 and its Alternatives*. London: New Economics Foundation.

People 1st (2017a) *Migrant Workers in the Hospitality and Tourism Sector and the Potential Impact of Labour Restrictions*. London: People 1st.

People 1st (2017b) *Migration and Labour Restrictions/Brexit and the Visitor Economy*. London: People 1st.

Perkins, A. (2018) Green groups reject Gove's plan for the environment. *The Guardian* 11 May, p. 17.

Perraudin, F. and Solomon, K. (2017) Brexit may upset British bid to represent European culture. *The Guardian* 28 October, p. 20.

Pettifor, A. (2017) Brexit and its consequences. *Globalizations* 14 (1), 127–132.

Phillips, J. (2015) A quantitative-based evaluation of the environmental impact and sustainability of a proposed onshore wind farm in the United Kingdom. *Renewable and Sustainable Energy Reviews* 49, 1261–1270.

Phillips, T. (2016) Chinese indulge in post-Brexit shopping sprees as pound sinks. *The Guardian* 4 July, p. 36.

Pickett, G. (2016) *Flying Through the Clouds. Preparing for Uncertainty Following Brexit*. London: Deloitte.

Pike, A., Dawley, S. and Tomaney, J. (2010) Resilience, adaptation and adaptability. *Cambridge Journal of Regions, Economy and Society* doi:10.1093/cjres/rsq001: 1–12.

Pinder, J. and Usherwood, S. (2013) *The European Union: A Very Short Introduction*. Oxford, UK: Oxford University Press, 3rd. edn.

Pinsent Masons (2018) *Into the Breach: the Role of General Counsel in Navigating a Successful Brexit*. London: Pinsent Masons.

Polak, P.R. (2017) Brexit: Theresa May's red lines get tangled in her red tape. A commentary on the White Paper. *European Papers* 2 (1), 403–410.

Pope, C. (2018) 'Brexit busting' ferry launched from Dublin Port. *Irish Times*, 20 April. www.irishtimes.com/news/ireland/irish-news/brexit-busting-ferry-launched-from-dublin-port-1.3468760 (accessed 17 November 2018).

Portes, J. (2018) *Too High a Price? The Cost of Brexit – What the Public Thinks*. London: Global Future. http://ourglobalfuture.com/reports/too-high-a-price-the-cost-of-brexit-what-the-public-thinks/ (accessed 23 April 2018).

Powell, J. (2017) Brexit Britain has rendered itself irrelevant. *The Guardian* 13 November, p. 25.

Pratt, G. and Rosner, V. (2012) Introduction: the global and the intimate. In G. Pratt and V. Rosner (eds) *The Global and the Intimate: Feminism in Our Time* (pp. 1–27). New York, NY: Columbia University Press.

Prazeres, L. and Findlay, A. (2017) *An Audit of International Students' Mobility to the UK*. Southampton, UK: ESRC Centre for Population Change, Working Paper 82.

Press Association (2017) UK emigrants permitted to stay in Spain in event of no-deal Brexit. *The Guardian* 22 October. www.theguardian.com/politics/2017/oct/22/uk-emigrants-spain-no-deal-brexit-eu (accessed 23 October 2017).

Press Association (2019) More than 400,000 EU citizens apply to stay in UK. *The Guardian* 11 April, p. 10.

Preston, A. (2018) If we want to shape the landscape that's fit for all, we must stop romanticising about it. *The Guardian* 14 January, p. 35.

Prime Minister's Office (2017) PM unveils plans for a modern Industrial Strategy fit for Global Britain. London: Press release. www.gov.uk/government/news/pm-unveils-plans-for-a-modern-industrial-strategy-fit-for-global-britain (accessed 31 October 2017).

Prime Minister's Office (2018) PM speech on our future economic partnership with the European Union. London: Press release, 2 March. www.gov.uk/government/speeches/pm-speech-on-our-future-economic-partnership-with-the-european-union (accessed 24 April 2018).

Quinn, B. and Doward, J. (2017) Special rights for Irish citizens in UK 'at risk'. *The Observer* 3 December, p. 6.

QS (2019) *University Rankings*. London: Quacquarelli Symonds. www.topuniversities.com/university-rankings (accessed 20 June 2019).

Radjenovic, A. (2018) *European Travel Information and Authorisation System (ETIAS)*. Brussels: European Parliamentary Research Service PE 599.298, 18 October.

Ramachandran, S. (2006) Visiting friends and relatives (VFR) market: A conceptual framework. TEAM *Journal of Hospitality & Tourism* 3 (1), 1–10.

Ramsay, A. (2018) The High Court found that Vote Leave broke the law in a new way. *Open Democracy UK*, 14 September. www.opendemocracy.net/uk/brexitinc/adam-ramsay/high-court-found-that-vote-leave-broke-law-in-different-way# (accessed 20 September 2018).

Rankin, J. (2018) The remainers: Europe united as it learns lessons of UK's departure. *The Guardian* 23 March, p. 19.

Rankin, J. (2019) EU climate action proposals condemned as 'meaningless'. *The Guardian* 11 June, p. 23.

Reckless Agency (2017) *What Impact Could Brexit Really Have on UK Tourism?* London: Reckless Agency.

Reid, C.T. (2016) Brexit and the future of UK environmental law. *Journal of Energy and Natural Resources Law* 34 (4), 407–415.

Reinhardt, U. (2011) Closer, shorter, cheaper: how sustainable is this trend? In R. Conrady and M. Buck (eds) *Trends and Issues in Global Tourism 2011* (pp. 27–36). Berlin/Heidelberg: Springer.

Reiser, D. (2003) Globalisation: An old phenomenon that needs to be rediscovered for tourism? *Tourism and Hospitality Research* 4, 306–320.

Ren, C. and Blichfeldt, B. (2010) One clear image? Challenging simplicity in place branding. *Scandinavian Journal of Hospitality and Tourism* 11 (4), 416–434.

Resolution Foundation (2018) *The RF Earnings Outlook: Quarterly Briefing Q1 2018*. London: Resolution Foundation.

Richards, G. and Rotariu, I. (2015) Developing the eventful city in Sibiu, Romania. *International Journal of Tourism Cities* 1, 89–102.

Richards, G. and Wilson, J. (2004) The impact of cultural events on city image: Rotterdam, cultural capital of Europe 2001. *Urban Studies* 41, 1931–1951.

Rita, P. (2000) Tourism in the European Union. *International Journal of Contemporary Hospitality Management* 12 (7), 434–436.

Roberts, D. (2017) Brexit: British officials drop 'cake and eat it' approach to negotiations. *The Guardian* 3 July. www.theguardian.com/politics/2017/jul/02/british-officials-drop-cake-and-eat-it-approach-to-brexit-negotiations (accessed 15 May 2018).

Roberts, D. and Rankin, J. (2017) Blue passports mean red tape, say EU officials. *The Guardian* 23 December, pp. 1, 7.

Roberts, L. and Hall, D. (2001) *Rural Tourism and Recreation: Principles to Practice*. Wallingford: CABI Publishing.

Robertshaw, P. (2017) Scotland in Europe update 24th November 2017. www.alynsmith.eu/scotland_in_europe_update_24th_november_2017?utm_campaign=

sie171124_2&utm_medium=email&utm_source=alynsmith (accessed 25 November 2017).

RTE (Radio Telefís Eireann) (2016) Brexit protesters set up mock checkpoint on border. Dublin: RTE 8 October. www.rte.ie/news/2016/1008/822475-border-rallies-brexit/ (accessed 6 May 2018).

RTE (Radio Telefís Eireann) (2017) Stormont protest over Brexit warns against hard border. Dublin: RTE 29 March. www.rte.ie/news/brexit/2017/0329/863595-brexit-border/ (accessed 6 May 2018).

Rudd, K. (2019) Think the Commonwealth can save Brexit Britain? No way. *The Guardian* 12 March, Journal p. 4.

Saarinen, J., Rogerson, C.M. and Hall, C.M. (2017) Geographies of tourism development and planning. *Tourism Geographies* 19 (3), 307–317.

Saayman, M. and Saayman, A. (2004) Economic impact of cultural events. *South African Journal of Economic and Management Sciences* 7, 629–641.

Savage, M. (2017) May's social mobility tsar quits with attack on 'fairness' failure. *The Observer* 3 December, pp. 1, 9.

Savage, M. (2018a) Brexit could wreck green agenda, says UN. *The Observer* 19 May, p. 18.

Savage, M. (2018b) Just how hostile has Britain become? *The Observer* 22 April, pp. 17–19.

Sayer, A. (2014) *Why We Can't Afford the Rich*. Bristol: Policy Press.

Schlanger, N. (2017) Brexit in betwixt. Some European conjectures on its predictability and implications. *The Historic Environment: Policy and Practice* online at: https://doi.org/ 10.1080/17567505.2017.1358324.

Seaton, A., and Tie, C. (2015) Are relatives friends? Disaggregating VFR travel 1994– 2014. In E. Backer and B. King (eds) *VFR Travel Research: International Perspectives* (pp. 28–45). Bristol: Channel View Publications.

Sedgley, D., Haven-Tang, C.L. and and Cockburn-Wooten, C. (2017) reactions to and anticipated consequences of Brexit for UK older people with second homes in Spain. *Critical Tourism Studies Proceedings*, Article 22. https://digitalcommons.library.tru. ca/cts-proceedings/vol2017/iss1/22/ (accessed 18 September 2018).

Settle, M. (2017) Tensions grow over open Irish border. *The Herald* (Glasgow) 22 November, p. 6.

Shani, A. and Uriely, M. (2012) VFR tourism: The host experience. *Annals of Tourism Research* 39 (1), 421–440.

Sima, C. (2017) Brexit impacts on British tourism. *Strategii Manageriale* (Piteşti, Romania) 10 (5)[34], 297–303.

Slawson, N. and Crerar, P. (2018) Number of Britons seeking citizenship elsewhere in EU doubles. *The Guardian* 10 April, p. 5.

Smeral, E. (2009) The impact of the financial and economic crisis on European tourism. *Journal of Travel Research* 48, 3–13.

Smeral, E. (2010) Impacts of the world recession and economic crisis on tourism: Forecasts and potential risks. *Journal of Travel Research* 49, 31–38.

Sørenson, A. and Nilsson, P.Å. (1999) Virtual rurality versus rural reality in rural tourism – contemplating the attraction of the rural. 8th. Alta, Norway: Nordic Symposium in Hospitality and Tourism Research, 18–21 November.

Sparrow, A. (2018) Lidington pledges to hand 'vast majority' of EU powers to devolved administrations. *The Guardian* 26 February, pp. 8–9.

Steinbach, J. (1995) River related tourism in Europe - an overview. *GeoJournal* 35 (4), 443–458.

Steiner, C., Richter, T., Dörry, S., Neisen, V. *et al.* (2013) *Economic Crisis, International Tourism Decline and its Impact on the Poor*. Madrid: United Nations World Tourism Organization.

Stone, J. (2016) EU students applying to British universities plummet by 9 per cent following Brexit vote. *The Independent* 27 October. www.independent.co.uk/news/uk/politics/ eu-students-applying-to-british-universities-plummet-by-9-percent-following-brexit-vote-a7382516.html (accessed 10 January 2018).

Stothard, M. (2018) UK and Spain reach Brexit deal on Gibraltar. *Financial Times* 19 October. www.ft.com/content/6fc997b6-d379-11e8-a9f2-7574db66bcd5 (accessed 24 October 2018).

Subramanian, S. (2017) How to sell a city: The booming business of place branding. *The Guardian* 7 November, pp. 27–29.

Summers, H. (2017) Gibraltar: Spain drops sovereignty link to EU deal. *The Guardian* 9 August, p. 9.

Sun, J. (2012) *Japan and China as Charm Rivals: Soft Power in Regional Diplomacy*. Ann Arbor MN: University of Michigan Press.

Swain, H. (2018) Stop Brexit group 'The kids don't want it being forced on them'. *The Guardian* 20 February, p. 35.

Sýkora, L. and Bouzarovski, S. (2012) Multiple transformations: Conceptualising the post-communist urban transition. *Urban Studies* 49 (1), 43–60.

Taraves, J. and Jorge, R.P. (2016) The path to Brexit: the view from Portugal. In C. Wyplosz (ed.) *What to do with the UK? EU Perspectives on Brexit* (pp. 107–116). London: CEPR Press.

Taylor, R., Shanka, T. and Pope, J. (2004) Investigating the significance of VFR visits to international students. *Journal of Marketing for Higher Education* 14 (1), 61–77.

Tebbit, N. (1990) Being British, what it means to me: time we learned to be insular. *The Field* (London) 272, 76–78.

TFEU (2012) Consolidated Version of the Treaty on the Functioning of the European Union. *Official Journal of the European Union* C 326/60, 26 October, 344pp. https://eur-lex.europa.eu/legal-content/EN/TXT/?uri=CELEX%3A12012E%2FTXT (accessed 22 April 2018).

Tham Min-En, A. (2006) Travel stimulated by international students in Australia. *International Journal of Tourism Research* 8 (6), 451–468.

the3million (2017) the3million: Defending citizensrights [sic] at the European Parliament. www.the3million.org.uk (accessed 21 February 2018).

Thunberg, G. (2019) *No One is Too Small to Make a Difference*. London: Penguin Random House UK.

Tisdall, S. (2018) China's pitiless war on Muslim Uighurs poses dilemma for the west. *The Observer* 16 September, p. 28.

TMF Group (2018) *The Financial Complexity Index 2018*. Amsterdam: TMF Group. www.tmf-group.com/en/news-insights/publications/2018/financial-complexity-index/ (accessed 7 October 2018).

Toly, N. (2017) Brexit, global cities, and the future world order. *Globalizations* 14 (1), 142–149.

Tourism Alliance (2016) *EU Referendum. Impacts on Tourism*. London: The Tourism Alliance. www.tourismalliance.com/downloads/TA_A_391_416.pdf (accessed 16 December 2017).

Tourism Alliance (2017) *Tourism After Brexit 2017: A Post-Brexit Policy Agenda for the UK Tourism Industry*. London: The Tourism Alliance. www.tourismalliance.com/downloads/TA_A_391_419.pdf (accessed 12 August 2017).

Tourism Ireland (2017) Island of Ireland Overseas Tourism Performance: 2016 Facts and Figures. Dublin/Coleraine: Tourism Ireland. www.tourismireland.com/TourismIreland/media/Tourism-Ireland/Press Releases/Press Releases 2017/Facts-and-Figures-2016.pdf?ext=.pdf (accessed 8 May 2018).

Tourism Ireland (2018a) Organisational Structure. Dublin/Coleraine: Tourism Ireland. www.tourismireland.com/About-Us/Organisation-Structure (accessed 6 May 2018).

Tourism Ireland (2018b) Tourism Ireland SOAR (Situation & Outlook Analysis Report) March 2018. Dublin/Coleraine: Tourism Ireland. www.tourismireland.com/TourismIreland/media/Tourism-Ireland/Press Releases/Press Releases 2017/SOAR-March-2018.pdf?ext=.pdf (accessed 8 May 2018).

Tourism Ireland (2018c) Tourism Ireland welcomes growth of +7% in overseas visitors for January-March 2018. Dublin/Coleraine: Tourism Ireland, press release 26 April. www.

tourismireland.com/Press-Releases/2018/April/Tourism-Ireland-welcomes-growth-of-7-in-overseas (accessed 6 May 2018).

Townsend, M. (2017) Thatcher aide derides blue passport 'nostalgia'. *The Guardian* 24 December, p. 4.

Travelsupermarket.com (2017) What could Brexit mean for our holidays? www.travelsupermarket.com/en-gb/blog/travel-advice/what-could-a-brexit-mean-for-our-holidays/ (accessed 18 March 2018).

Travelzoo (2016) European tourists to the UK could drop by a third following Brexit: Travelzoo survey reveals impact to tourism in a post-Brexit Britain. https://press.travelzoo.com/european-tourists-to-the-uk-could-drop-by-a-third-following-brexit/ (accessed 25 November 2017).

Travis, A. (2017) Irish border was crossed 110m times last year. *The Guardian* 22 September, p. 8.

Tribe, J. (2008) Tourism: A critical business. *Journal of Travel Research* 46 (3), 245–255.

Trudgill, P. (2017) Could vote stay have stuck in our heads better than remain? *The New European* 23 June, p. 31.

Turner, C. (2019) Trade war with US driving Chinese students to Britain, study reveals. *Daily Telegraph* 28 May.www.telegraph.co.uk/news/2019/05/28/trade-war-us-driving-chinese-students-britain-study-reveals/ (accessed 31 May 2019).

Uberoi, E. (2015) *European Union Referendum Bill 2015–16*. London: House of Commons Library Briefing Paper 07212. www.researchbriefings.files.parliament.uk/documents/CBP-7212/CBP-7212.pdf (accessed 22 October 2017).

UCL (University College, London) (2016) Mandarin Excellence Programme. London: UCL Institute of Education. https://ci.ioe.ac.uk/mandarin-excellence-programme/ (accessed 8 November 2018).

UK Government (2017a) *Building Our Industrial Strategy*. London: Green Paper.

UK Government (2017b) *The United Kingdom's Exit From and New Partnership with the European Union*. London: Stationery Office Cm 9417.

UK Government (2018a) *A Green Future: Our 25 Year Plan to Improve the Environment*. London: Stationery Office.

UK Government (2018b) *Erasmus+ in the UK if There's No Brexit Deal*. London: UK Government, 23 August.

UK Government (2018c) *Mobile Roaming if There's No Brexit Deal*. London: UK Government, 13 September.

UK Government (2018d) *The Future Relationship between the United Kingdom and the European Union*. London: HMSO Cm 9593.

UK Government (2018e) *The UK's Future Skills-based Immigration System*. London: HMSO, Cm 9722.

UK Government (2018f) *Travelling Within the Common Travel Area and the Associated Rights of British and Irish Citizens if There is No Brexit Deal*. London: UK Government, 13 September.

UK Government (2019) £1.6 billion Stronger Towns Fund launched. London: UK Government, 4 March.

UKCCC (Committee on Climate Change) (2018) *Implications of the Vote to Leave the EU. London: Committee on Climate Change*. www.theccc.org.uk/tackling-climate-change/the-legal-landscape/implications-of-vote-to-leave-eu/ (accessed 4 November 2018).

UK2070 Commission (2019) *The First Report of the UK2070 Commission: Fairer and Stronger – Rebalancing the UK Economy*. Sheffield, UK: UK2070 Commission.

UNCTAD (United Nations Conference on Trade and Development) (2018) *World Investment Report 2018*. Geneva: UNCTAD.

Universities UK (2017) *The Economic Impact of International Students*. London: Universities UK. www.universitiesuk.ac.uk/policy-and-analysis/reports/Pages/briefing-economic-impact-of-international-students.aspx (accessed 20 January 2018).

Universities UK (2018) *Brexit and UK Universities.* London: Universities UK www.universitiesuk.ac.uk/policy-and-analysis/brexit/Pages/brexit-and-universities.aspx (accessed 24 October 2018).

UNWTO (United Nations World Tourism Organization) (2014) *Tourism Highlights 2014.* Madrid: UNWTO.

UNWTO (United Nations World Tourism Organization) (2017a) *2016 Annual Report.* Madrid: UNWTO.

UNWTO (United Nations World Tourism Organisation) (2017b) *European Union Short-Term Tourism Trends.* Madrid: UNWTO.

UNWTO (United Nations World Tourism Organization) (2018) *European Union Short-Term Tourism Trends Volume 2 2018–1.* Madrid: UNWTO. www.e-unwto.org/doi/book/10.18111/9789284419593 (accessed 3 April 2018).

Valentino, P. (2018) L'Italia in panchina nella partita di Merkel e Macron. *Corriere Della Sera Online* (Milan), 16 March.www.corriere.it/opinioni/18_marzo_17/italia-panchina-partita-merkel-macron-15d17794-294a-11e8-b8d8-0332a0f60590.shtml (accessed 15 May 2018).

Varoufakis, Y. (2017) *And the Weak Suffer What They Must? Europe, Austerity and the Threat to Global Stability.* London: Vintage.

Vaughan, A. (2018) Plastic bag loophole ends but what about coffee cups? *The Guardian* 12 January, pp. 6–7.

Viggo, S. (2017) UK: Chinese student recruitment weaker than expected. *The Pie News* (London) 27 October. https://thepienews.com/news/uks-chinese-student-recruitment-poorer-than-expected/ (accessed 20 December 2017).

Viña, G. (2016) Third of foreign students less likely to come to UK after Brexit. *The Financial Times* 28 July. www.ft.com/content/c179cb10-53f3-11e6-9664-e0bdc13c3bef (accessed 10 December 2017).

VisitBritain (2013) *Great China Welcome.* London: VisitBritain.

VisitBritain (2015) *Annual review 2014/15.* London: VisitBritain.

VisitBritain (2016a) *Annual Review 2015/16.* London: VisitBritain.

VisitBritain (2016b) *VisitBritain Launches New Campaigns With Expedia Across France, Germany and US To Drive Inbound Tourism.* London: VisitBritain.

VisitBritain (2017a) *Great Britain Domestic Overnight trips Summary – All trip purposes – 2017.* London: VisitBritain.

VisitBritain (2017b) *Market and Trade Profile China.* London: VisitBritain.

VisitBritain (2018a) *Inbound Town Data.* London: VisitBritain.

VisitBritain (2018b) *Regional Spread of Inbound Tourism.* London: VisitBritain.

VisitScotland (2016) *Tourism Development Framework for Scotland: Role of the Planning System in Delivering the Visitor Economy.* Edinburgh: VisitScotland.

Visser, G. and Ferreira, S. (2013) Tourism and crisis: A never-ending story? In G. Visser and S. Ferreira. (eds) *Tourism and Crisis* (pp. 1–11). London: Routledge.

Vrbaski, L. (2016) Flying into the unknown: The UK's air transport relations with the European Union and third countries following 'Brexit'. *Air and Space Law* 41 (6), 421–444.

Wahl, P. (2017) Between Eutopia and nationalism: A third way for the future of the EU. *Globalizations* 14 (1), 157–163.

Walker, P. (2018) No 10 defends passport maker as number one for cyber security. *The Guardian* 24 March, p. 21.

Walker, P. and Mason, R. (2017) Davis admits 'impact forecasts don't exist' but escapes censure. *The Guardian* 7 December, pp. 14–15.

Walker, P. and Triest, V. (2019) Embattled PM hopes EU membership will see off 'curse of history'. *The Guardian* 11 June, p. 29.

Wang, M.J. (2017) UK group wants Chinese students to stay after studies end. *China Daily* 28 November. www.chinadaily.com.cn/a/201711/28/WS5a276cc8a3107865316d3df6.html (accessed 3 December 2017).

Ward, S. (1997) Anglo-Commonwealth relations and EEC membership: The problem of the Old Dominions. In G. Wilkes (ed.) *Britain's Failure to Enter the European Community 1961–63* (pp. 93–107). London: Routledge.

Ward, S. (2001) A matter of preference: The EEC and the erosion of the old Commonwealth relationship. In A. May (ed.) *Britain, the Commonwealth and Europe* (pp. 156–180). Basingstoke: Palgrave Macmillan.

Warrell, H. (2017) Falling pound cuts tuition fees for foreign students by up to 20%. *The Financial Times* 20 April. www.ft.com/content/a1489f3e-25e3-11e7-a34a-538b4cb30025 (accessed 18 March 2018).

Weaver, D.B. (2003) The contribution of international students to tourism beyond the core educational experience: Evidence from Australia. *Tourism Review International* 7 (2), 95–105.

Weaver, D.B. (2010) Geopolitical dimensions of sustainable tourism. *Tourism Recreation Research* 35 (1), 47–53.

Wei, S. (2017) Parents fear UK classes losing character due to many Chinese students. *Global Times* (Beijing) 17 November. www.globaltimes.cn/content/953252.shtml (accessed 18 March 2018).

Wellings, B. and Baxendale, H. (2015) Euroscepticism and the Anglosphere: Traditions and dilemmas in contemporary English nationalism. *Journal of Common Market Studies* 53 (1), 123–139.

Welsh Government (2016) *Partnership for Growth: Strategy for Tourism 2013–2020. Strategy Progress Review*. Cardiff: Welsh Government.

West, K. (1994) Britain, the Commonwealth and the global economy. *The Round Table* 83 (332), 407–417.

Weston, R., Davies, N., Scuttari, A., Wagner, M. and Pechlaner, H. (2014) *The Cost of Non-Europe in the Single Market in Transport and Tourism: III-Tourism Policy and Passenger Rights*. Brussels: European Parliamentary Research Service (EPRS), European Union. www.europarl.europa.eu/RegData/.../2014/.../EPRS_STU(2014)510988_REV1_EN.pdf (accessed 22 April 2018).

White, M. (2017) May's EU aid package. *The New European* 26 October, pp. 8–10.

White, M. (2018) All sides keep powder dry. *The New European* 22 March, pp. 14–15.

Whyte, P. (2018) What the latest agreement means for transport and tourism. *Skift* 28 November. https://skift.com/2018/11/28/what-the-latest-brexit-agreement-means-for-travel-and-tourism/ (accessed 19 December 2018).

Wilkinson, A. (2017) Generation Brex. *The New European* 10 August, p. 11.

Williams, A.M. and Hall, C.M. (2002) Tourism, migration, circulation and mobility: The contingencies of time and place. In C.M. Hall and A.M. Williams (eds) *Tourism and Migration: New Relationships Between Production and Consumption* (pp. 1–52). Dordrecht: Kluwer.

Willis, G. (2017) Brexit is a chance to save our small farms. *The Guardian* 6 December, p. 32.

Wilson, T.M. (2016) The Brexit vote in regard to Northern Ireland and Ireland (Forum). *Social Anthropology* 24 (4), 498–499.

WNS (2018a) Decisionpoint: Brexit: Travel and Leisure Industry. Mumbai, India: WNS Holdings. www.wnsdecisionpoint.com/brexit/gclid/eaiaiqobchmiqttxpvpn2–givcjgbch2viqcyeaayasaaegksifd_bwe#resp-tab3 (accessed 22 April 2018).

WNS (2018b) Decisionpoint: UK Aviation: No Open Skies Post Brexit. Mumbai, India: WNS Holdings. www.wnsdecisionpoint.com/brexit/gclid/eaiaiqobchmiqttxpvpn-2givcjgbch2viqcyeaayasaaegksifd_bwe#resp-tab3 (accessed 22 April 2018).

Worldometers (2019) China population (LIVE). Worldometers. www.worldometers.info/world-population/china-population/ (accessed 15 June 2019).

Wong, E.P.Y, Mistilis, N. and Dwyer, L. (2011a) A framework for analysing intergovernmental collaboration – the case of ASEAN tourism. *Tourism Management* 32 (2), 367–376.

Wong, E.P.Y, Mistilis, N. and Dwyer, L. (2011b) A model of ASEAN collaboration in tourism. *Annals of Tourism Research* 38 (8), 882–899.

Wood, C. and Busby, E. (2018) A-level results day 2018: Chinese overtakes German among students for first time, amid decline in languages. *The Independent Online* www.independent.co.uk/news/education/education-news/a-level-results-day-chinese-german-modern-foreign-languages-decline-students-a8494156.html (accessed 18 August 2018).

Wood, Z. and Kollewe, J. (2019) Tesco and M&S build up stocks of tinned food. *The Guardian* 11 January, p. 7.

Worstall, T. (2018) Global Future's Brexit scenarios are wrong over WTO tariffs. *Continental Telegraph* 18 April (accessed 23 April 2018).

Worth, O. (2017) Reviving Hayek's dream. *Globalizations* 14 (1), 104–109.

Wright, B. (2016) There's a sinister strain of anti-intellectualism to Gove's dismissal of 'experts'. *The Daily Telegraph* 21 June. www.telegraph.co.uk/business/2016/06/21/in-defence-of-experts-whether-they-support-leave-or-remain/ (accessed 17 October 2017).

WTM (World Travel Market) (2017) Brexit complexity confirmed by Brits' range of travel worries. London: WTM, 6 November. https://news.wtm.com/tag/industry-report-2017/ (accessed 17 November 2018).

WTTC (World Travel and Tourism Council) (2017) *Travel & Tourism Economic Impact*. London: WTTC. www.wttc.org/-/media/files/reports/economic-impact-research/regions-2017/world2017.pdf (accessed 25 March 2018).

Zhang, A., Zhong, L., Xu, Y., Wang, H. and Dang, L. (2015) Tourists' perception of haze pollution and the potential impacts on travel: Reshaping the features of tourism seasonality in Beijing, China. *Sustainability* 7 (3), 2397–2414.

Zhao, X.Y. (2017) Britain still appeals to students, despite Brexit. *China Daily* 18 January. www.chinadaily.com.cn/newsrepublic/2017-01/18/content_27993382.htm (accessed 25 March 2017).

Index